From the Metal Up

Fundamentals of Programmable Computers

Matthew Blagden

Owl Island Software

Published By
Owl Island Software
www.owlisland.com

ISBN 978-0-9812800-0-4

Owl Island Software also publishes books in electronic format.

Contents

Part III: Software Programming 205

Introduction

Computer programming is a valuable skill that virtually anyone can acquire. In recent years, programming has become a popular profession in both corporate and entrepreneurial environments. Aside from providing career opportunities, casual programming is a satisfying endeavor. Requiring no costly materials or great expanses of space, it can make an ideal hobby.

Software development is also a very practical skill. Even casual computer users can become frustrated with fruitless searches for programs that perform specific tasks, finding either expensive solutions or no solutions at all. The ability to write your own software allows you to create programs that perform exactly the tasks you want, in the manner that you prefer.

Even without great amounts of time dedicated to designing your own programs, a thorough understanding of how programs work can aid everyday computer use. Knowledge of the inner workings of software systems often provides insight into solving confusing computer problems.

Why This Book Was Written

Computers have evolved over many generations of hardware and software. As hardware grew more powerful and complex, new programs were built on top of older ones. As the programs were extended, they became more abstract and generic. Many modern programming languages are now taught with virtually no references to the actual hardware and software they run upon. As a result, many textbooks simply rely on a variety of metaphors and abstractions to describe the underlying layers of the system.

Although ignoring details of how the system functions can lead to a faster start, a thorough understanding of the entire system can lead to higher proficiency in the long term. High-level programs cannot run without interacting with other software and hardware, and an encounter with the outside system is inevitable.

This book aims to give an overview of each layer of the system, starting at the hardware and finishing with an introduction to the C# programming language. The presented concepts are applicable to a myriad of programming topics that range far beyond the situations and languages explicitly addressed in the examples.

Who Should Read This Book

This book is intended for anyone with an interest in computers or computer programming. No previous experience in computer programming is required. Concepts will be explained without the use of any advanced mathematical or electrical terminology. The most advanced mathematical concept in this book is the use of exponents. As some software does need to be installed to use the provided examples, knowledge of basic computer use is also required.

Those with prior programming experience may still find this book useful for understanding the evolution of computers, and how each layer of a computer system affects the programs they write. Knowledge of the environment in which software runs is an invaluable asset when addressing complicated programming issues.

What You Need to Use This Book

Programming is best learned by practicing and experimenting. To get the most out of this book you should use the sample programs and try changing them to suit your needs. To make use of all examples in the book you will need to run several programs that can be obtained at no charge from the accompanying resources website and from Microsoft®. Specifically, you will need a computer running the Microsoft Windows® operating system, as well as the following software:

- The accompanying resources bundle
- The Express Edition of Microsoft Visual C#®

Links to both of these software packages may be found on the accompanying resources page at http://www.owlisland.com/books.

The resources bundle also contains copies of example programs shown in this book.

Part I: Hardware

1

Automation

1.1 Making Life Easier

People have been using machines to make life easier for a great deal of recorded history. Physical work was the first application of mechanical assistance. Archimedes began the progression as far back as the 3rd century BC with a variety of simple machines that could ease the burden of physical labor. As mathematics played a significant role in many mechanical inventions, it was a natural progression to try to automate calculations as well.

Automating logical tasks proved to be significantly more complicated than their physical counterpart. It is quite easy to observe repetitive labor in the physical world and imagine how a powered machine could manipulate objects in a similar manner. Unfortunately, the way a brain manages to perform mental calculations and logical reasoning is quite difficult to observe. A whole new class of machines would have to be invented to mimic the logical actions carried out by the human mind.

1.2 First Steps

Eventually, machines were devised that could perform basic calculations. The beginnings of logical automation were not the most elegant solutions, but were certainly a step in the right direction. In almost all cases, large racks of levers and gears interlocked to create a giant chain reaction when a crank was turned. After a sufficient amount of cranking, the complex interaction of the mechanical parts would calculate the result. The resulting values were often punched into paper logs or simply implied by the final position of the gears or levers.

Maintaining these machines was tedious, and the operation was relatively slow, but the costs were justified. Barring any mechanical failure, results were always perfectly accurate. Steel gears do not make mental math errors, they do not become tired, and they certainly do not get confused. The consistent results of automation that had been reaped in physical endeavors could now be enjoyed in computational tasks.

The sheer size and complexity of powerful versions of these calculators made them quite expensive. Only a select few organizations could obtain one of the more complex mathematical machines. Even once acquired, only trained individuals could possibly operate the intricate machines, and even greater skill was required to understand and repair them. The idea of a general-purpose calculating machine had been proven, but it was a long way from becoming readily available to the community at large. Complex computation was still quite a laborious task.

1.3 Moving Forward

As the initial problem of mapping physical gears to conceptual processes had been solved, these initial calculators were destined to come into common use. Clever designs reduced the space required to house the masses of levers and gears, pushing the machines closer to everyday users. Companies formed around these mechanical devices, bringing their benefits to all forms of repetitive calculation. Everything from mail sorting to payroll administration to cash register calculations could be replaced by specialized calculating machines. Workers could quickly actuate buttons and levers instead of performing tedious, error-prone calculations.

As the supply of these machines grew, so did the demand for faster and increasingly complex implementations. This demand revealed a major flaw in the first designs: the only way to increase the abilities of these machines was to add an exponentially increasing amount of gears and levers. More moving parts created more potential points of failure and made production prohibitively expensive. The only way to avoid this problem was with a completely new design, brought on by two important changes: use of the transistor, and the binary number system.

2

Binary

2.1 The Numbers You Know

You are probably quite accustomed to seeing and understanding numbers every day without considering how they are composed. For the sake of learning a new number system, it is helpful to review the basics of how numbers work.

The numbering system you are probably most familiar with is the decimal number system. The number 875 is a valid decimal number consisting of three digits. Each digit has a value; the values are eight, seven, and five in this case. The name "decimal" originates from the use of ten possible values for each digit: zero through nine. This also led to the slightly more descriptive name, "base ten".

Each number is simply a set of consecutive digits. Each digit in a number can only be one of the values allowed by the numbering system, and any set of consecutive digits is a valid number in the numbering system. You do not need to have ever seen the number 18395 to know that it is a valid decimal (base ten) number. You can simply observe that each digit is one of the ten values allowed in the decimal number system.

Once you can recognize numbers, the next step is to understand their meaning. The number 7584 is composed of four decimal digits, and is simply a short way of writing:

$$(7 \times 1000) + (5 \times 100) + (8 \times 10) + (4 \times 1)$$

or:

$$(7 \times 10 \times 10 \times 10) + (5 \times 10 \times 10) + (8 \times 10) + (4 \times 1)$$

or:

$$(7 \times 10^3) + (5 \times 10^2) + (8 \times 10^1) + (4 \times 10^0)$$

As you can see, the last digit in the number is a multiple of one, and each digit as you move to the left is multiplied by a factor ten times larger.

It is worth noting that adding zeroes on the right hand side of non-zero digits does make a difference. The number 500 is different from 5000 because the extra zero on the right-hand side moved the five into a different place (from the digit that is a multiple of 100 to the digit that is a multiple of 1000). Adding zeroes on the left hand side of all non-zero digits does not make a difference:

$$(6 \times 100) + (3 \times 10) + (9 \times 1)$$

$$(0 \times 10000) + (0 \times 1000) + (6 \times 100) + (3 \times 10) + (9 \times 1)$$

The two lines above are the expanded versions of 600 and 00600. If you carry out the math for each, you will see that they have the same value. No matter how many zeroes you add to the left of a number, it will remain unchanged. For this reason, any zeroes on the left side of non-zero digits are typically omitted (which is why 00600 may look quite odd).

All of these principles were observed using the decimal system, but can be applied to different numbering systems with bases other than ten.

2.2 Simpler Numbers

To automate calculations, physical objects must be created to represent and manipulate numbers. Simplicity is the key to reducing costs, material requirements, and design complexity. The basic task of adding two one-digit numbers is actually quite complex when using the decimal number system. There are eighty-one combinations of numbers to be added together (zero through nine matched with each of zero through nine), and eighteen possible results. Creating physical gears and levers for each of these can become expensive very quickly. For large scale computer design, a simpler number system would be quite advantageous.

Base ten obtained its name from having ten possible values for each digit. A "base one" number system would not make much sense as each digit could only have the one value allowed by the system: zero. Therefore, the simplest practical number system is "base two", providing a choice of two values for each digit: zero and one. Just as base ten had a common name – decimal – base two is often referred to by a simpler name: binary.

Following the same rules as were seen with decimal, binary values can be broken down into their components as well. The binary number 11001 is a short form of:

```
(1 × 2 × 2 × 2 × 2) + (1 × 2 × 2 × 2) + (0 × 2 × 2) + (0 × 2) + (1 × 1)
```

or:

$$(1 \times 2^4) + (1 \times 2^3) + (0 \times 2^2) + (0 \times 2^1) + (1 \times 2^0)$$

or:

```
(1 × 16) + (1 × 8) + (0 × 4) + (0 × 2) + (1 × 1)
```

The pattern is the same as it was in base ten, where there are ten possible digits, except that the number two is used instead of ten, to indicate that there are only two possible digits. The last digit is a multiple of one, and each digit to the left is multiplied by a factor two times larger. If you carry out the calculation, you will see that the binary number 11001 has the same value as the decimal number 25.

Many people learning decimal find it useful to think of each digit as a "place" or "column". In decimal, this breaks numbers down into digits in the "ones", "tens", "hundreds", and "thousands" columns etc. For example, 628 is six hundreds, two tens, and eight ones. A similar scheme can be devised for binary. Instead of ones, tens, hundreds, and thousands, binary digit places are ones, twos, fours, and eights etc. For example, the binary number 11001 is one sixteen, one eight, zero fours, zero twos, and one one.

You cannot always tell which base a number uses simply by looking at it. The number 101 is valid in both decimal and binary, but indicates a different quantity in each. To avoid explicitly stating "decimal" or "binary" before each number is written, numbers are often written with their base noted. The decimal number 25 may be written as "$(25)_{10}$" to signify that it is expressed in decimal (base ten). It may also be written as "11001b" or "$(11001)_2$" in binary (base two). Although these numbers look different, $(11001)_2$ and $(25)_{10}$ have the same value. Decimal and binary simply express values in different ways.

2.3 Counting

Counting works the same in binary as it does in decimal. Consider counting up from zero in decimal. The last digit in the number is increased by one until it reaches its maximum value. In decimal, the maximum value for any single digit is, of course, nine. Increasing the value again resets the digit to zero and the digit to its left is increased by one. Even if there is no digit to the left, there are always assumed to be "invisible" zeroes in all unwritten positions. If the digit to the left is also at its maximum, the digit on its left is incremented as well. This repeats until a digit can be incremented without passing its maximum. For an example, consider counting past 49999. Raising the rightmost digit by one causes it to reset to zero, and the digit to the left is incremented, resetting it to zero, which causes the digit to its left to be incremented, and so on, until finally the four can be incremented to five, yielding 50000.

Applying these rules to the binary number system reveals how binary counting works. The only difference between the two situations is that the maximum value for a digit is one instead of nine.

Decimal	Binary
0	0
1	1
2	10
3	11
4	100
5	101
6	110
7	111

2.4 Terminology

Most areas of study that grow to a substantial size eventually develop their own unique terminology. Many other disciplines are based on deciphering existing objects or phenomena, as would be done in biology or physics. Computers differ from these areas of study in that they are based on invention; computers were created to serve a specific task, they were not observed in the wild. Computer terminology is often created in an ad-hoc manner by the inventors or first users of the technology. As a result, many of the terms used in computing are quite literal, and rife with colloquialisms and puns. Binary number terminology is no exception.

The fundamental unit of digital computation is the bit. A bit is simply a single binary digit, a one or a zero. Numbers larger than one require additional bits, or digits, as seen earlier in this chapter. Just as a two-digit decimal number can only represent one hundred values (zero through ninety nine), a two-bit binary number can only represent four values (zero through three).

You may notice that binary numbers often have extraneous zeroes on the left-hand side. Although these digits do not change the amount represented by the number, they are commonly written when using binary to better represent how computers store values. Computers use physical parts to represent everything they do. If a computer has sufficient parts to store 32 binary digits, the number represented by those parts is always a 32-bit number. Values that are larger than 32 bits cannot be stored; there are not enough parts to store them. Values smaller than 32 bits can be stored, but there will still be 32 physical parts, so the "extra" digits to the left are set to zero. In general, a 32-bit computer manipulates 32-bit numbers, while a 64-bit computer uses numbers up to 64-bits wide.

This limit must always be kept in mind when performing calculations. Numbers used in computing often have all digits written, including preceding zeroes, to make this limit explicit. When the maximum size of a number is four bits, the number five would commonly be written 0101 in binary. In an eight-bit space, the same number would be written 00000101.

Naturally, as sizes are so important, names have been assigned to the various quantities of bits.

Bits	Name
1	bit
4	nibble / nybble / half-octet / quartet
8	byte / octet

Computer terminology also differs from normal use of language when referring to individual objects. In everyday use, numbering often begins at one. If person wanted to refer to each of ten items on a desk, they may assign them numbers one through ten. In computing, these items would likely be referred to as items zero through nine. This convention may seem unusual at first, but eventually becomes quite natural. As you progress through this book, the convenience of this numbering system should become obvious.

An example of this new numbering system can be seen when referring to individual bits. The four-bit number 0100 has four bits: bit zero on the right through bit three on the left. The only bit that has a value of one is bit number two. The leftmost bit in an eight-bit number is bit seven. The rightmost bit is bit zero. By numbering bits from right to left, bit number zero always represents the ones column, bit number two always represents the fours column, and so on, even when the size of a number changes.

The names for quantities of bytes are less standard than the names used for quantities of bits. The predominant use of binary in computers has led many quantities to be based upon powers of two. Unfortunately, the names commonly used for quantities based on powers of two are the same as the names used in quantities based on powers of ten.

Term	Number of Bytes
kilobyte (KB)	2^{10} or 10^3 (1 024 or 1 000)
megabyte (MB)	2^{20} or 10^6 (1 048 576 or 1 000 000)
gigabyte (GB)	2^{30} or 10^9 (1 073 741 824 or 1 000 000 000)
terabyte (TB)	2^{40} or 10^6 (1 099 511 627 776 or 1 000 000 000 000)

The advantages of using powers of two are quite clear when working in binary. Using units based on powers of two, one kilobyte is $(10000000000)_2$ bytes. Using units based on powers of ten, one kilobyte is $(1111101000)_2$ bytes. Of course, the units based on powers of ten are more convenient when working in base ten, making a kilobyte $(1000)_{10}$ bytes instead of the $(1024)_{10}$ bytes that result from using powers of two.

Materials related to programming will often use the units based on powers of two (1024 bytes per kilobyte) as they are more convenient for programmers. Advertising materials often use powers of ten (1000 bytes per kilobyte) because smaller units lead to larger capacity claims. A music player than can hold ten million bytes of music can be advertised as having a 10 megabyte capacity using units based on powers of ten ($10000000 / 10^6$ = 10), or only 9.5 megabytes if using units based on powers of two ($10000000 / 2^{20}$ = 9.5).

A new set of terms has been developed to specify quantities of bytes. Rather than using the ambiguous "kilobyte", "megabyte", "gigabyte", and "terabyte" units, some people choose to use "kibibyte", "mebibyte", "gibibyte", and "tebibyte". These terms always refer to the quantities based on powers of two. The new terms also have new abbreviations: KiB, MiB, GiB, and TiB. Which terms you use is your choice. The newer terms are unambiguous, but are also much less common.

2.5 Mathematics

The basic rules of mathematics also apply to each number system. As a reminder of the exact steps for adding numbers, carry out the following addition:

$$\begin{array}{r} (236)_{10} \\ (184)_{10} \\ \hline \end{array}$$

The sum is determined one column of digits at a time, beginning with the rightmost digits:

$$\begin{array}{r} 1 \\ (239)_{10} \\ (184)_{10} \\ \hline 3 \end{array}$$

Adding the rightmost digits nine and four gives $(13)_{10}$. The last digit of the number (three) gets written as the last digit of the sum, while the rest of the number (one) gets carried to the next column. The next digit of the sum is now determined using the three, the eight, and the one that was carried in from the previous step:

$$
\begin{array}{r}
\mathit{11} \\
(239)_{10} \\
(184)_{10} \\
\hline
23
\end{array}
$$

This yields $(12)_{10}$. Again, the last digit (two) is written as the next digit in the sum, and the rest (one) is carried to the next column. The final digit is calculated using the two, the one, and the one that was carried in from the previous step:

$$
\begin{array}{r}
\mathit{11} \\
(239)_{10} \\
(184)_{10} \\
\hline
(423)_{10}
\end{array}
$$

The exact same technique can be applied to binary numbers as well. Consider the following sum:

$$
\begin{array}{r}
(0011)_2 \\
(0111)_2 \\
\hline
\end{array}
$$

Again, starting with the rightmost digits, a one on the top and a one on the bottom:

$$
\begin{array}{r}
\mathit{1} \\
(0011)_2 \\
(0111)_2 \\
\hline
0
\end{array}
$$

Adding gives $(10)_2$. This makes zero the last digit of the sum. After carrying the rest (the one) we can move on to the next column, which is now a one on the top, a one on the bottom, and a one carried in from the previous step:

$$
\begin{array}{r}
\mathit{11} \\
(0011)_2 \\
(0111)_2 \\
\hline
10
\end{array}
$$

Adding the three ones yields $(11)_2$, making the next digit in the sum a one, and causing another one to be carried. The next digit is determined by a zero on the top, a one on the bottom, and a one carried in:

$$\begin{array}{r} \mathit{111} \\ (0011)_2 \\ (0111)_2 \\ \hline 010 \end{array}$$

This gives another $(10)_2$. Again, the zero is written as the next digit in the sum, and the one is carried. The final digit is calculated using a zero on the top, zero on the bottom, and a carried-in one from the previous step:

$$\begin{array}{r} \mathit{111} \\ (0011)_2 \\ (0111)_2 \\ \hline (1010)_2 \end{array}$$

According to this math: $(0011)_2$ plus $(0111)_2$ equals $(1010)_2$. You can confirm this by converting each number to decimal. $(0011)_2$ is $(3)_{10}$, $(0111)_2$ is $(7)_{10}$, and these together make $(10)_{10}$. It appears that the calculation was correct, as $(10)_{10}$ is $(1010)_2$ (that is, one eight, and one two).

The binary math shown above used four-bit numbers. In a four-bit system all computations are trimmed to four bits. Adding $(1111)_2$ and $(0100)_2$ on paper would yield $(10011)_2$. If this calculation was done in a computer that could not manipulate more than four bits, the result would be truncated to four bits, giving a final result of $(0011)_2$.

2.6 Negativity

So far we have only examined positive values. Making a decimal number negative is quite simple: write a "–" before any number and it is deemed to be negative. Unfortunately, this simple notation causes quite a bit of complexity later. Negativity is expressed using a special symbol that is not a valid digit, and care must be taken to account for negative signs whenever any computation is performed.

Minimizing complexity is one of the major goals of using the binary number representation. Rather than using a simple representation and complex math, we can use a slightly more complicated representation and greatly simplify computations. Negative binary numbers are commonly represented using a format called two's complement. The "two's complement" of a binary number can be found by inverting each bit (changing ones to zeroes and zeroes to ones), and then adding one.

For example, consider the number five in a four-bit system. Five would be written as $(0101)_2$. Inverting the bits would yield $(1010)_2$. Finally, adding one gives the representation of the negative of five: $(1011)_2$. One of the benefits of two's complement is that binary addition remains unchanged. You can see this by adding seven and negative five:

$$\begin{array}{r} 111 \\ (0111)_2 \\ (1011)_2 \\ \hline (0010)_2 \end{array}$$

Note that because we are working in a four-bit system only the rightmost four bits of the result are actually calculated (the leftmost column of the calculation does result in a carried one, but it is simply discarded). The calculation above shows that using the same method of binary addition yields the correct result (seven plus negative five gives two). No consideration needs to be given to whether either of the numbers being added is negative. Correct results can also be observed when the sum is negative:

$$\begin{array}{r} 11 \\ (0011)_2 \\ (1011)_2 \\ \hline (1110)_2 \end{array}$$

This calculation shows the addition of three and negative five. The correct answer is negative two. The binary form of positive two is $(0010)_2$. Inverting the bits gives $(1101)_2$. Adding one gives the final representation of negative two: $(1110)_2$. It appears the calculation still worked, giving the correct result of negative two.

You may have noticed that the binary number $(1011)_2$ could be interpreted as either negative five or positive eleven. The interpretation of values is up to you; the math will work either way. If you are treating numbers as "signed" (possibly negative), then the leftmost bit will indicate whether the number is positive or negative. This makes the previous equation represent the addition of three and negative five, giving a result of negative two. If you treat numbers as "unsigned" (always positive), then the previous equation represents the addition of three and eleven, giving a result of positive fourteen. It is up to you to remember whether you are treating numbers as normal (unsigned) or two's complement (signed) values. So long as you interpret all numbers as signed, or all numbers as unsigned, the addition will make perfect sense.

2.7 Keeping it Concise

Binary is as simple as number systems can get, but is also quite verbose. It would be nice if there were a more convenient way to write these numbers. The computer will use binary underneath, so a system at least somewhat close to binary is useful.

Decimal is quite familiar to people, but it is not very desirable for communicating with computers. The number of bits required to represent any given number is very important, given the limited length of numbers in computers. The number of digits in a decimal number does not map to a number of bits very well. Even a two-digit decimal number can need anywhere from four to seven bits:

$$(10)_{10} = (1010)_2 \qquad\qquad (99)_{10} = (1100011)_2$$

More importantly, changing the last decimal digit affects almost all of the binary digits, as does changing the first one. There is no clear correlation between each decimal digit and each of the binary digits:

$$(64)_{10} = (100000)_2 \qquad (63)_{10} = (011111)_2 \qquad (54)_{10} = (0110110)_2$$

As with many numeric problems in computing, using powers of two can simplify matters. One of the most common number systems used in programming is the hexadecimal number system. Hexadecimal is base sixteen, so sixteen values are possible for each digit: 0, 1, 2, 3, 4, 5, 6, 7, 8, 9, A, B, C, D, E or F. Hexadecimal, commonly referred to as "hex", has several notations. The hexadecimal number 3A6C is written as 0x3A6C, 3A6Ch, or $(3A6C)_{16}$. 0x3A6C is the short way of writing:

$$(3 \times 16 \times 16 \times 16) + (10 \times 16 \times 16) + (6 \times 16) + (12 \times 1)$$

or:

$$(3 \times 16^3) + (10 \times 16^2) + (6 \times 16^1) + (12 \times 16^0)$$

or:

$$(3 \times 4096) + (10 \times 256) + (6 \times 16) + (12 \times 1)$$

Hexadecimal and binary are much more closely correlated. Each hexadecimal digit maps to exactly four bits, with $(0)_{16}$ equal to $(0000)_2$, and $(F)_{16}$ equal to $(1111)_2$. Any eight-digit hexadecimal number can be expressed in 64 bits, any four-digit hexadecimal number will fit in 32 bits, and any two-digit hexadecimal number will fit in 8 bits (one byte).

Changing any one digit of a hexadecimal number will only ever change the four bits that it maps to:

$$(CF6)_{16} = (1100\ 1111\ 0110)_2 \qquad (C06)_{16} = (1100\ 0000\ 0110)_2$$

Converting between hexadecimal and binary is also much easier than converting between decimal and binary. Because each hexadecimal digit affects exactly four binary digits, numbers can be converted four bits at a time (which means you never need to convert a binary number larger than sixteen). For example, consider the number $(1101111010010100110)_2$. Manually converting this number to decimal would be quite tedious. In decimal it is the sum:

$$2^{18} + 2^{17} + 2^{15} + 2^{14} + 2^{13} + 2^{12} + 2^{10} + 2^7 + 2^5 + 2^2 + 2^1$$

or:

$$262144 + 131072 + 32768 + 16384 + 8192 + 4096 + 1024 + 128 + 32 + 4 + 2$$

Converting to hexadecimal is much easier. Simply break the binary number into four-bit segments (remember, adding zeroes on the left does not change the value):

$$0110\ 1111\ 0100\ 1010\ 0110$$

Next, convert each four-bit segment into its corresponding hexadecimal digit:

$$(0110)_2 = (6)_{16}$$

$$(1111)_2 = (F)_{16}$$

$$(0100)_2 = (4)_{16}$$

$$(1010)_2 = (A)_{16}$$

$$(0110)_2 = (6)_{16}$$

Concatenate the digits and you will have the hexadecimal answer: 0x6F4A6. Converting back to binary is simply the reverse (convert each hexadecimal digit to four bits). Hexadecimal provides an easy way to write numbers concisely, with strong correlation to the binary version of the number that the computer will use.

2.8 Exercises

1. Convert the following decimal numbers into binary: 15, 7, 127, 63. Why do they look the way they do in binary?

2. Add the following pairs of binary numbers: (1101 + 0011), (1010 + 0101), (1111 + 0001). Assume a four-bit system.

3. In regular decimal numbering you can write positive zero and negative zero as "0" and "-0". Can you make a negative zero in two's complement notation? If so, what happens when you add it to another non-zero number? Assume a four-bit system.

4. When using signed (possibly negative) binary numbers, you can tell if a number is negative by looking at its leftmost bit. How can you tell if a hexadecimal number is negative?

3

The Transistor

3.1 Flow Control

The overall goal of creating computers is to have physical machines perform tedious logic and calculations. We have decided to use the binary number system because of its simplicity. Unfortunately, there are not actually "ones" and "zeroes" in nature that we can manipulate; we need to choose something physical and use it to represent different values.

The original computing machines used the position of gears and levers to indicate values. For large-scale computing, we will need something that is much smaller and capable of faster movement. The choices are plentiful. Air pressure, water pressure, and light intensity are all valid options for representing different values. Of course, electrical voltage is a popular choice. Older mechanical computers made use of electricity to move their many physical parts, but we can actually use electricity to represent the numbers themselves.

You may have noticed that the choices for a physical medium to represent values all seem to be things that flow, as opposed to rotate like gears or tilt like levers. In fact, the basic idea behind modern computing machines is to interconnect a large network of devices that control flow. You are probably already familiar with several flow-control devices: faucets control the flow of water, valves control the flow of air, and light switches control the flow of electricity.

The important difference between these devices and the switches used in computing is that these household devices cannot readily interact. To automate tasks, the switches need to be able to control each other, rather than requiring manual manipulation. To be specific, we need a valve that not only controls the flow of some substance, but is actuated by that same type of substance. This can be accomplished with virtually any substance we choose, but a water-based system is one of easiest to observe:

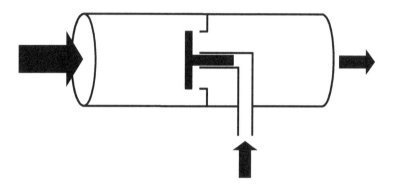

These valves have three important parts: a source, output, and input (or "control"). The diagram shown above has the source entering on the left, output exiting on the right, and the input being actuated at the bottom. The large arrow indicates that high pressure water is entering the source on the left-hand side. The small arrow at the bottom indicates that only low-pressure water is entering the control part, causing the valve to open slightly. The net result is that very little water can pass through the valve, causing the output to be low-pressure water. Changing the input to high pressure yields the following:

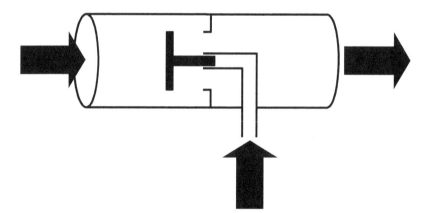

With high pressure applied to the input, the valve opens much wider and allows water to pass freely. Because high-pressure water is entering on the left, high-pressure water can exit on the right.

This leads to the next pair of scenarios, in which only low-pressure water is entering the input:

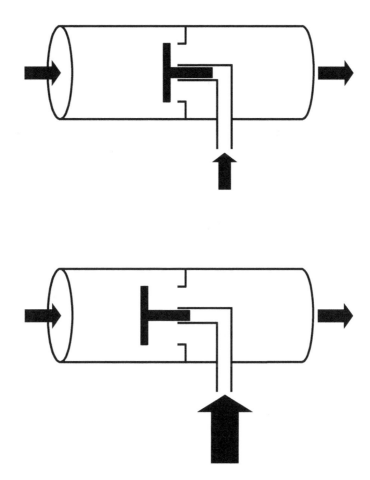

Regardless of how far the control valve is opened, if the water entering the pipe source is at low pressure, only low-pressure will be observed on the output side.

3.2 Electronic Valves

The examples presented in the previous section illustrate how the flow of some matter can be controlled using the same type of matter. Although entirely possible, we do not commonly create computers that operate using pipes of water or tubes of air; using electricity proves to be much more convenient. The same behaviors that were exhibited by the water valves can be created using an electronic device called a transistor. The symbol for a transistor looks very different than a water pipe, but has the same important parts: source on the left, output on the right, and input (or "control") at the bottom.

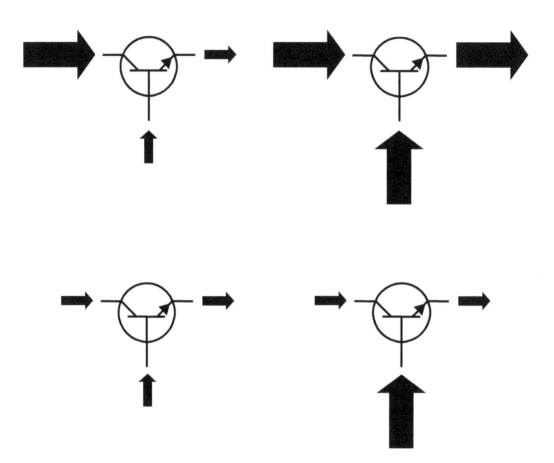

These diagrams use wiring notation. A straight thin line represents a wire, as can been seen at the input, output, and control parts of the transistor. Crossed wires are not connected unless joined by a solid dot. Of the two pairs of wires shown below, only the pair on the right is connected:

All connected wires carry the same voltage. If high voltage is applied to a wire, all connected wires also carry high voltage. If low voltage is applied to a wire, all connected wires will carry low voltage. If both high voltage and low voltage are applied to connected wires, the wires will all carry high voltage. This operation is similar to how water pipes (of the same size) would behave. If high-pressure water is fed into a pipe, all connected pipes will also carry high-pressure water. If low pressure water is fed into a pipe, all connected pipes will contain low-pressure water. If both high-pressure and low-pressure water are fed into connected pipes, the water inside the pipes will be at a high pressure. The pressure (or voltage) will only change when a pipe (or wire) reaches a valve (or transistor). The pressure on the other side of the valve will be lower if the valve is not open very far. High pressures can only pass through if the valve is wide open.

3.3 Doing the Opposite

In all of the previous examples, a high input voltage allowed a high source voltage to flow to the output. We can also create the opposite effect. In the water-based example, the valve can be made to close, rather than open, when actuated.

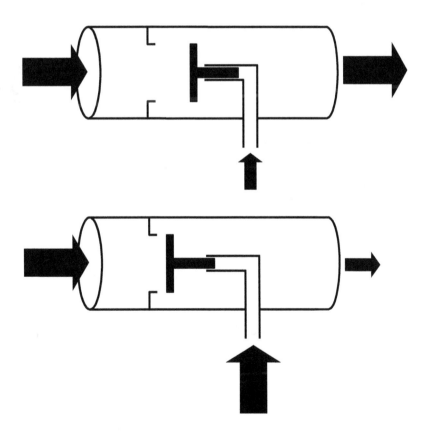

This same concept can be applied to create the opposite type of transistor. The "opposite" is indicated by the small circle on the bottom.

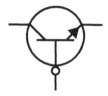

As illustrated with the water pipes, this transistor's behavior is the inverse of the first transistor we examined.

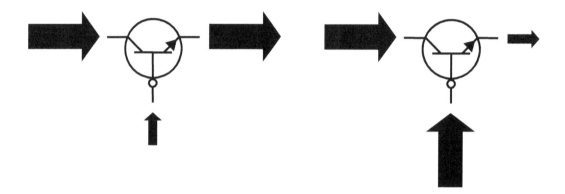

When the control is high voltage, the "valve" is pushed nearly shut, causing the output to be low voltage. When the control is low voltage, the valve can open completely, allowing the high voltage to flow freely to the output. The output is the opposite of the control. This type of transistor can be referred to as an inverter.

3.4 Exercises

1. Draw a pair of transistors with a single input wire connected to both sources and a single output wire connected to both outputs.

2. Examine the device created in question 1. If the source is connected to high voltage, when is the output high voltage?

3. Draw a pair of transistors with the output of one transistor connected to the source of the other.

4. Examine the device created in question 3. If the source is connected to high voltage, when is the output high voltage?

4

Logic Gates

4.1 Grouping Transistors

The only reason that we are using transistors is because we chose to use electrical voltages to represent values. Because we are simply assigning numbers to physical phenomena, how we assign them is up to us. The binary number system only has two values – one and zero – so we only need to make two assignments. This book will use higher voltage electricity to represent one and lower voltage electricity to represent zero. If a "one" is put onto an input wire, it actually means higher voltage electricity is present within the wire. If a device is outputting a "zero" on a particular wire, that wire has lower voltage electricity within it. Three wires with high voltage, low voltage, and high voltage, respectively, would represent the binary value $(101)_2$, which could also be written as the decimal value $(5)_{10}$.

"High voltage" may conjure mental images of bright signs warning of danger. In the context of computing, "high voltage" simply refers to a voltage that is higher than what is referred to as "low voltage". It is not actually important how high or low the voltages are, only that one is higher than the other. A computer could be designed to use 0.1 volts and 0.5 volts as "low voltage" and "high voltage", or 50 volts and 1000 volts as "low voltage" and "high voltage". Indeed, there are a wide variety of transistors designed to operate on very different ranges of voltage.

We want to use computers for logical deduction in addition to calculation. We can assign logical values to voltages just as we assigned numerical values for the binary number system. We can call high voltage "true" and low voltage "false". If a device outputs "true", it is actually placing high voltage on a wire (which we also use to represent the value "one"). If the output of a device is "false", it is actually outputting a low-voltage signal. As you can see, the physical implementation is simply varying voltages; what we use these voltages to represent is completely up to us. Electronic machines can automate computation and logic as long as they manipulate voltages in the same manner we wish to manipulate numerical values or logical values.

Transistors are both very useful and very basic devices. Common electronics use millions of transistors. To simplify the job of connecting all of these transistors together, we can group several transistors into devices capable of performing basic logic tasks. We can then use these basic logic units as the building blocks for more complicated devices. These logical building blocks are called "gates". The following sections will define the basic gates we need, and then show how they can be implemented using transistors.

4.2 NOT

The first gate we will examine is the inverter, or "NOT" gate. This gate has only two (significant) connections, input and output, and is drawn as follows:

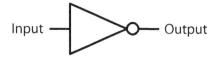

The desired output is the inverse of the input. Gate behavior is often summarized in a simple table.

Input	Output
false / 0	true / 1
true / 1	false / 0

The inverter is supposed to output a value opposite to what it receives on its input. If we are dealing with logic, this means "true" is converted to "false" and "false" is converted to "true". If we are dealing with numbers, this means "1" is converted to "0", and "0" is converted to "1". Of course, all of these values are simply assignments we made to various voltages. To obtain the desired behavior, this device will have to accept one voltage (high or low), and output the other.

For subsequent gates we will only deal with the logical values - "true" and "false" – but keep in mind that you could also use the device to manipulate the numbers "one" and "zero". The gates operate on physical voltages; what we use voltages to represent is up to us.

The inverter is the simplest gate, and can be made with just a single inverting transistor. A more detailed explanation of how this inverter functions can be seen in the previous chapter.

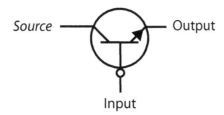

The transistor diagram has thee connections, while the inverter gate symbol only has two. Every gate is directly connected to the high-voltage power source. This connection is labeled "source" on the transistor diagram and is always high voltage. Because these connections are not directly related to the logic that the gate performs, they are simply omitted from the gate-level diagrams. These connections are required to make the transistors function, and are actually present in physical implementations of gates. Although omitted on gate-level diagrams, they will be shown on transistor-level diagrams as dotted lines for the sake of clarity.

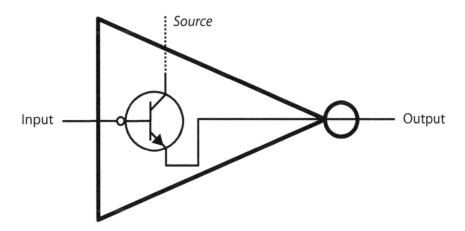

With the "source" connection hidden, all that is visible is the input and output. When the input is low voltage, the "valve" can open, allowing the high-voltage source to flow to the output. When the input is high voltage, the valve closes and allows only low-voltage electricity to reach the output. Note that this behavior is exhibited because this gate is built using an inverting transistor.

4.3 AND

The second gate we will design is called an "AND" gate. This device accepts two inputs, unlike the inverter. Note that the two inputs are simply numbered zero and one; this does not imply that they carry the values zero or one.

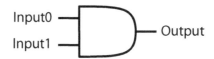

The function of this gate is to output true when both the first input is true AND the second input is true. A simple application of this gate could be a device to deduce if it is snowing outside. One input could be connected to a temperature sensor that indicates if it is cold outside. The other input could be connected to a moisture sensor that indicates if it is precipitating outside. The output of the device would indicate if it is snowing.

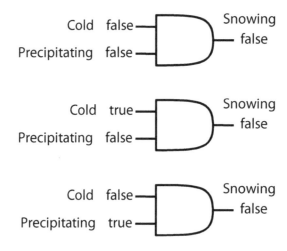

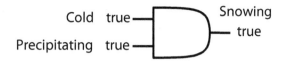

The precise behavior is summarized in the following table.

Input0	Input1	Output
false	false	false
true	false	false
false	true	false
true	true	true

The implementation of this gate requires two regular transistors.

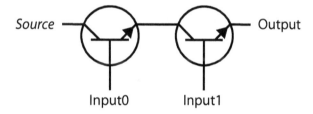

As in all gates, the "source" connection on the left is directly connected to the high-voltage source, and is always at a high voltage. As these are normal transistors, both transistors remain closed when neither of the inputs are high voltage, producing a low voltage at the output. Even if one of the two inputs is high voltage, opening one of the transistors, the other transistor will remain nearly closed. Only in the case when both inputs are high voltage will the high-voltage source be allowed to pass to the output uninhibited.

We can now fit this transistor implementation into the shape of the AND gate shown above.

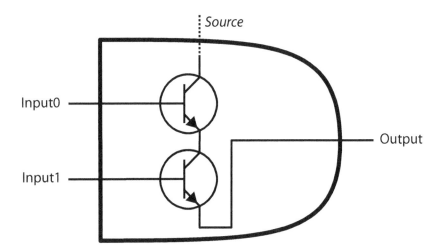

A gate that does the opposite of the AND gate is also quite useful. This device is referred to as a NAND gate. The symbol for a NAND gate is simply the symbol for an AND gate with a circle on the output, indicating the output is the opposite of what the AND gate would produce.

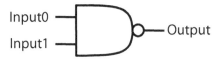

The desired behavior for the NAND gate is exactly the inverse of the AND gate: it outputs true whenever it is not the case that both inputs are true.

Input0	Input1	Output
false	false	true
true	false	true
false	true	true
true	true	false

A logical deduction that could use a NAND gate is deciding if a room is lit based on if the light bulb is burnt out and if it is night time. The room is lit when it is NOT the case that both the light bulb is burnt out AND it is night time.

Of course, simply attaching an inverter to the output of the AND gate would produce the desired result, but this would require an extra transistor. By changing the transistor type and layout we can also implement the NAND gate without using any extra transistors.

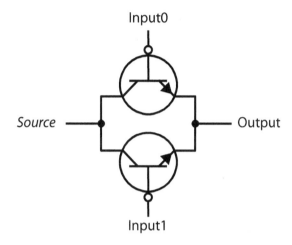

The first thing to notice is that we are now using inverting transistors, which remain open unless their input is high voltage. The high-voltage source also has two paths to the output. In this schematic, the only way the high-voltage source cannot reach the output is if both transistors are closed (that is, both inputs are high voltage). In true/false terms, the output is always true unless both inputs are true to close both paths between the source and output. This is exactly the desired behavior for a NAND gate.

We can fit this schematic into the NAND symbol as follows:

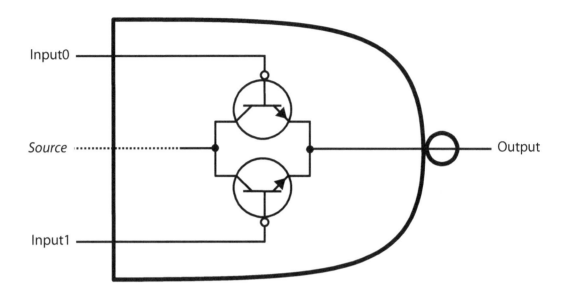

4.4 OR

The next gate we will need is the OR gate. The OR gate symbol is similar to the AND gate, but with a curved input side and pointed output side.

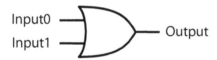

As its name implies, the output of this device is true whenever either of the inputs are true. The output is also true if both inputs are true.

A real-world example of a use for an OR gate could be deciding if a laptop power light should be turned on. With one input connected to the main circuit board and the other input connected to the battery charging circuit, the power light would be lit whenever the laptop is turned on or charging its battery (and, of course, if it is both in use and charging).

The following table describes the desired behavior.

Input0	Input1	Output
false	false	false
true	false	true
false	true	true
true	true	true

The AND gate required both transistors to be open to allow the high-voltage source to reach the output. To obtain the desired OR gate behavior, we must allow the source voltage to reach the output if either transistor is open.

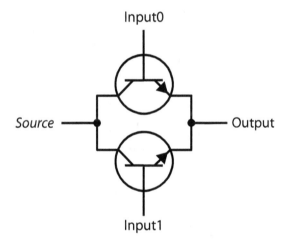

Using the transistor layout above, if either (or both) of the inputs are high voltage, an open path will be created from the high-voltage source to the output.

Fitting this schematic into the OR gate symbol yields the following:

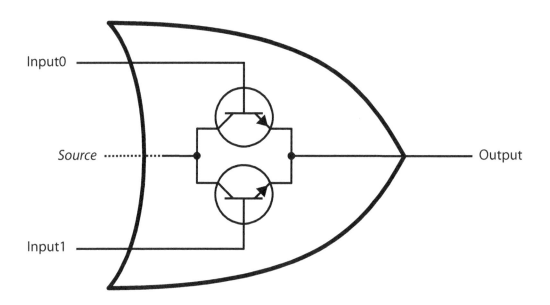

The opposite of the OR gate – the NOR gate – is also a useful gate to have.

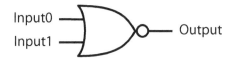

The behavior of the NOR gate is simply the opposite of the OR gate. The NOR gate detects the case when neither Input0 NOR Input1 is true. A NOR gate could be used to activate a lock on a security door. One input could be connected to a key-card reader on the outside of the building, and the other input connected to a motion sensor on the inside of the building. When there is neither a valid key-card on the outside, nor a person moving (exiting) from the inside, the lock should be engaged.

Input0	Input1	Output
false	false	true
true	false	false
false	true	false
true	true	false

Once again, we can create opposite behavior by changing the layout and type of transistors used:

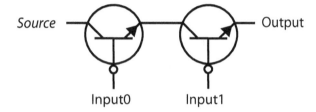

This layout causes the transistors to remain open when the inputs are low voltage. If either input is high voltage, its respective transistor will close, blocking the one and only path between the high-voltage source and the output.

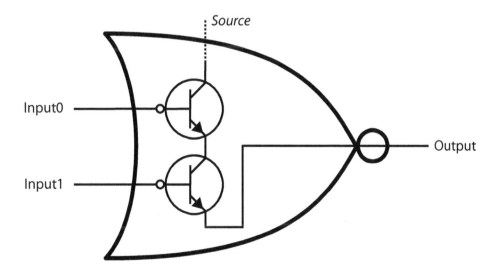

37

4.5 XOR

The final gate we will examine is the "exclusive or" gate, called the XOR gate.

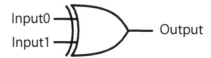

Recall that the output of the OR gate is true whenever at least one of the inputs are true. The XOR gate differs in that its output should be only be true when exactly one of its inputs is true, not both.

Input0	Input1	Output
false	false	false
true	false	true
false	true	true
true	true	false

One example of a XOR gate in use is a single light with two light switches. For the sake of convenience, two switches may be installed to control a single light (at either end of a staircase perhaps, so the user can control the light from either floor). Using an OR gate would require both switches to be off to turn off the light. Similarly, an AND gate would require both switches be set to on to turn on the light.

The desired behavior is to have the state of the light toggled whenever either switch is toggled. Starting with both switches off and the light off, toggling either switch will turn on the light. Now that the light is on, toggling either switch will turn it off (either causing both switches to be off, or both switches to be on). This is exactly the behavior of the XOR gate.

The XOR gate can be implemented using the previous gates we have already created, rather than using transistors directly. Writing a sentence describing the desired output often reveals how a gate-level schematic can be created. In the case of the XOR gate, we could write the following:

The output is true when:

- it is the case that Input0 is true or Input1 is true

  and

- it is not the case that Input0 is true and Input1 is true.

You can probably spot a few key words in the description.

The output is true when:

- it is the case that Input0 is true OR Input1 is true

  AND

- it is NOT the case that Input0 is true AND Input1 is true.

At this point, we can start using some gates to represent parts of the sentence.

"Input0 is true OR Input1 is true" is a clear use of an OR gate:

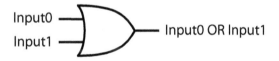

"Input0 AND Input1" implies an AND gate:

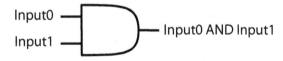

The AND gate will output true when Input0 is true AND Input1 is true. Looking back at the sentence, we can see that we want to detect when it is NOT the case that both Input0 is true AND Input1 is true:

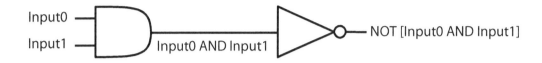

Although this solution is completely valid, we can create the same behavior with fewer parts:

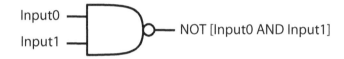

Looking back at the sentence one last time we can see that we have an OR gate to determine when the first point is true, and a NAND gate to determine when the second point is true. The output of the XOR gate should be true when both the first point AND second point are true. This calls for, of course, an AND gate.

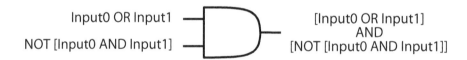

Substituting in the OR and NAND gate we used earlier completes the schematic for the XOR gate:

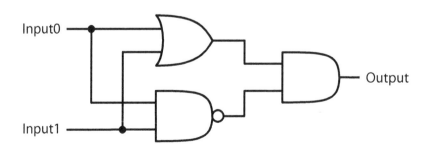

These interconnected gates make up the inner workings of the XOR gate.

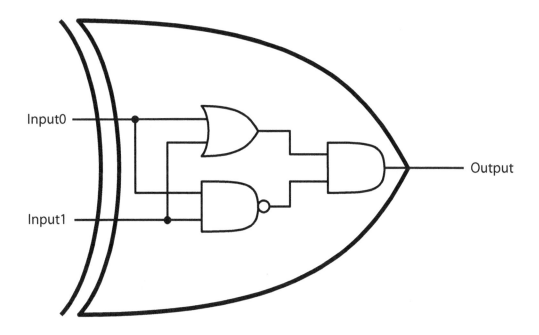

4.6 Exercises

Using the gates from this chapter:

1. Create the following gates using only NAND gates:

    a. NOT

    b. AND

    c. OR

    d. NOR

    What does this reveal?

2. Create a device with one input and one output that outputs true if the input has ever been true.

3. Design a device that outputs true when at least one of four inputs is true.

4. Design a device that accepts a three-bit number (three inputs, each representing one bit of the number), and outputs true when the input number is a one, three, four, or six.

5

Simple Machines

5.1 Memory

All of the circuits we have designed thus far have simply accepted input and generated output; once the input signal was removed, the output would also return to its previous state. We will need devices to temporarily "remember" values to carry out any calculations that require multiple steps. A clever arrangement of a few gates can create "memory" and actually store values.

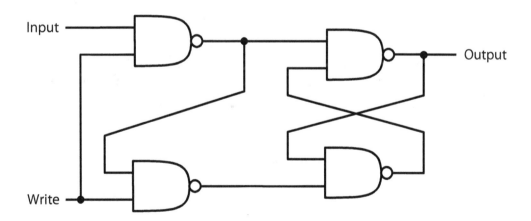

It is almost certainly not obvious why the above schematic stores values. The first step to understanding how this device saves values is to actually understand how it is intended to function.

To be able to store a value, we need to be able to differentiate between when the value at the input is the value we want to store and when it should simply be ignored. The "Write" signal indicates that the value on the input line should be stored. Once a value is stored, it will continue to appear on the "Output" line even after the input signal has changed.

To observe how this actually works, we can step through the operation with various values. Remember that because we are just assigning values to voltages, "1" and "true" are interchangeable (they are both just high voltage), and "0" and "false" are interchangeable (they are both just low voltage). We will use "true" and "false" to describe if we are writing, and "1" and "0" to describe the values we are writing. In the following images, a thin line has low voltage on it and a thick line is carrying high voltage.

Tracing through the behavior of a device is fairly straightforward. Simply set the inputs to the desired value, and then continue to resolve inconsistencies until the output of each gate is correct in relation to its inputs. To start off, we will store a zero. Storing a zero is accomplished by setting the input to the desired value (zero / low voltage) and enabling the write line (true / high voltage):

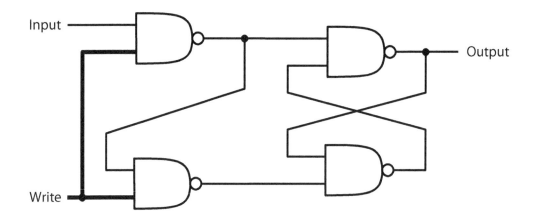

We can now examine how the gates behave, one step at a time. The two NAND gates on the left-hand side each have one input set to true / 1 / high voltage, and one input set to false / 0 / low voltage. As these are NAND gates, they detect when it is not the case that both inputs are true. This is indeed true for the left-hand gates, so their output is should be "true".

Setting the output of the two left-hand gates to "true" yields the following:

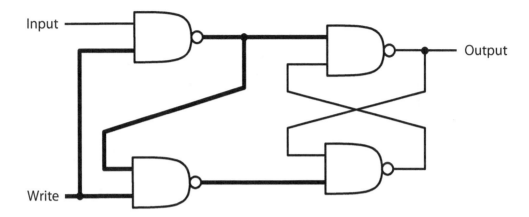

This change, in turn, causes more changes. We will start with the right-hand gates. Again, it is not the case that both inputs are true, so their output is true. We can also see that the bottom-left NAND gate has two true inputs, changing its output to false.

Making these three changes yields the following:

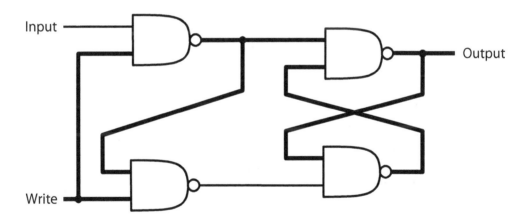

The only remaining gate with an incorrect output is the top-right NAND gate. As both inputs are true, the top-right gate should be outputting false.

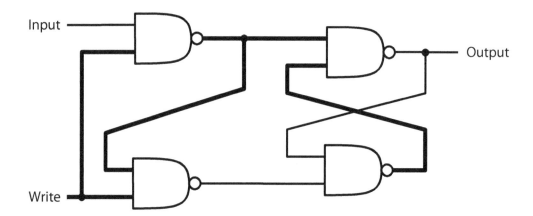

Each output now matches its inputs and the device is at a steady state. We were trying to store a zero, and we are indeed outputting a zero. The final step is to disable the "Write" line, now that we are done writing.

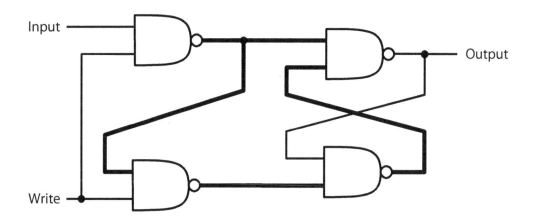

The bottom-left gate now outputs true, and nothing else is changed. The device is still outputting our stored zero. Just to ensure it is accomplishing what we desire, we can try changing the input to a one.

Changing the input to "one" gives:

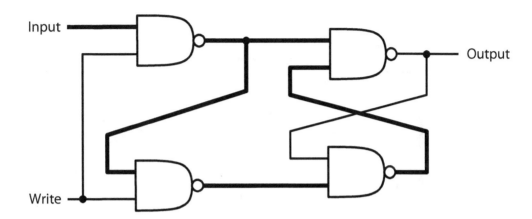

As hoped, changing the input had no effect on the output, as the device is not being told to store the input ("write" is false). Regardless of what value is set on the input, the output is kept at our stored value. We can now try storing a one. The input line is already set to "one", so we should only have to enable the write line.

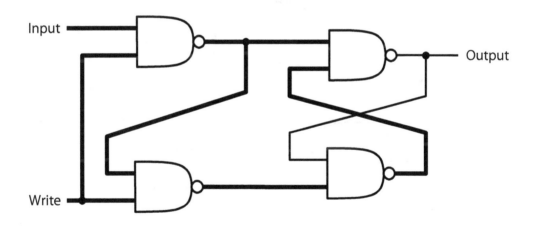

The outputs of the two left-hand NAND gates now change to false, as both its inputs are true.

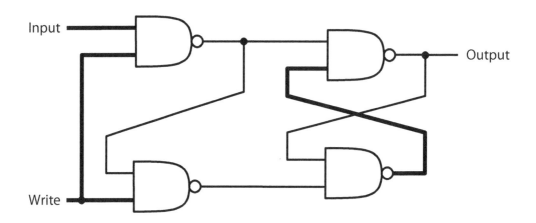

The top-right and bottom-left gates now need to change:

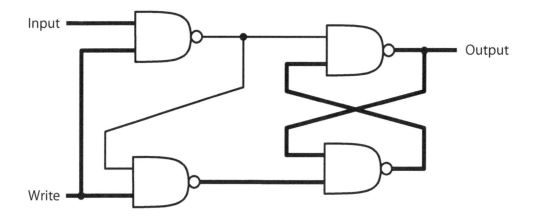

The bottom-right gate is now affected:

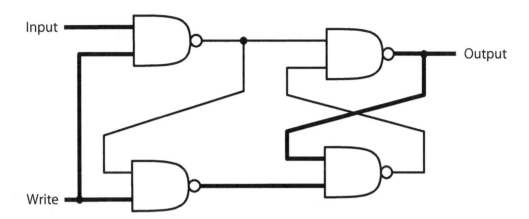

We have now satisfied the requirements of each gate. Now that we have finished storing the value "one", we can tell the device it should no longer be writing by setting the write line to false.

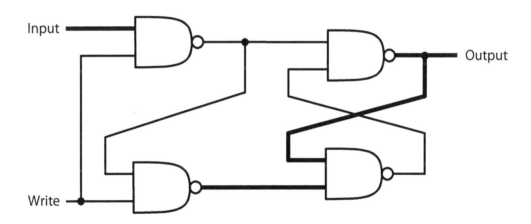

This affects the output of the top-left NAND gate.

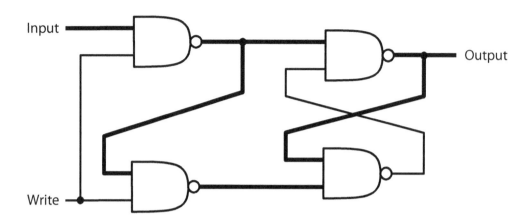

Once again, we are at a steady state. The device is properly outputting the "one" we stored. Just as one last check, we can try changing the input to a zero to ensure it does not affect the output.

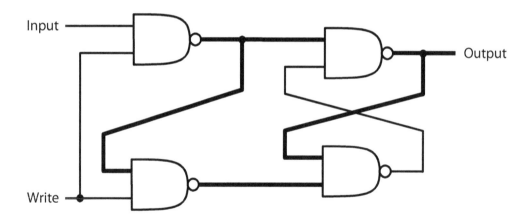

Absolutely nothing changes. We now have a device that can store one bit of information. Setting the "write" line to true captures the current input value and starts outputting that value. With the "write" signal changed back to false, no change in the input will affect the output.

5.2 Settling Time

Tracing through the behavior of the single-bit memory illustrates a very important point: once the desired input is applied to a device, the output does not change instantaneously. Gates can interact, and both the inter-gate connections and the device output can switch back and forth between various states before eventually settling to their final steady state.

The input signals must be held steady during the entire settling time to ensure the correct output is eventually generated. Changing the input part way through the settling shown in the previous section could change the eventual output. The amount of time required to reach the steady output is based on how many gates have to change values, and how complex their interactions are. This settling time is one of the major factors that determine how quickly calculations can be performed.

5.3 More Memory

Storing a single bit of data is a good start, but far from sufficient for most computing needs. Common computers manipulate 16-bit, 32-bit, and 64-bit data. A multi-bit storage device can be created simply by bundling together many one-bit storage devices. The first step is to place the one-bit storage device into a reusable package:

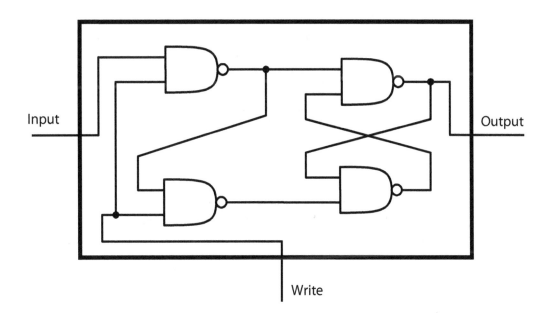

or simply:

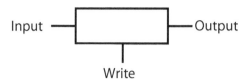

We will need four of the one-bit memory devices to store a four-bit number. Each bit of the number needs to be stored in a different location, but all four bits should be captured at the same time, to ensure the proper number is stored. This can be accomplished by connecting the write-enable lines of four parallel memory units.

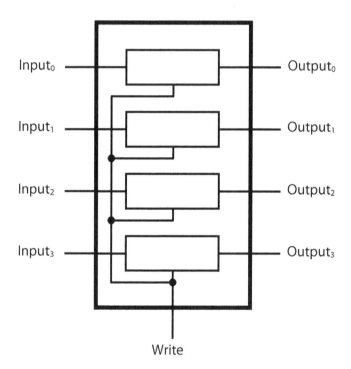

Because the four wires on the left make up a single four-bit number, they are all still called "Input" (not Input1 Input2 etc.), with subscripts appended to distinguish the individual bits. "$Input_1$" indicates bit one of the multi-bit signal named "Input".

This four-bit memory unit can be drawn as follows:

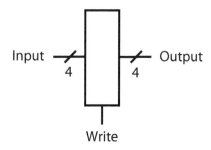

This schematic symbol uses some new notation. This notation groups all four input wires together, and all four output wires together. The slash across the line indicates that it is "wider" (contains more wires) than shown. A number may be written beside the slash to indicate the actual number of wires contained in the bundle. The write-enable line at the bottom does not have a slash as it is only a single wire.

The reason for this line grouping is to simplify the diagram. The gates we examined earlier had multiple input wires, but, because these inputs were unrelated, they were always drawn as separate wires. In the case of the four-bit memory unit, the four wires on the left are all conceptually part of the same input – the four-bit number to store. The same can be said of the four wires on the right; they are all a part of one conceptual output – the four-bit number that is currently stored.

Each of the wires in a bundle is physically independent and can carry a different value; they are only grouped for the sake of visual clarity. This notation groups wires that are all a part of the same multi-bit signal. Grouping some of the input wires and some of the output wires together would not be done, nor would a bundle of input or output wires be grouped with the write-enable wire.

We now have a device that can store a multi-bit number; this device is called a register. The pictures above only implement a four-bit register, but it should be clear that any number of one-bit memory units could be placed in a single package to store a value of any size.

If you saw the following symbol on a schematic, you could simply infer that it is an eight-bit register:

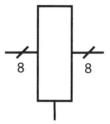

This device could be made of eight one-bit memory units, or two four-bit registers placed into a single package. How the eight-bit register is implemented is not relevant at the schematic level, so long as it stores and outputs an eight-bit value as intended.

5.4 Multiplexing

In addition to increasing the number of wires that make up a signal, we can also increase the number of signals connected to a device. Simply connecting multiple wires to the same input is not sufficient as the signals would mix. To have multiple input options available for a device, we need a way to select which one we want to use at any particular moment. The ability to choose between multiple inputs is known as "multiplexing".

Because only one input is to be active at a time, we need a unique number to identify each input. If our multiplexer chooses from ten input signals, we need to be able to input ten different numbers to specify which input we would like. Assuming we use the most obvious numbers, zero through nine, this implies that the "selector" input needs to be at least four bits wide (three bits can only represent zero through seven).

Whenever a multi-bit device is designed, it is simplest to start with the one-bit case. A one bit number can only specify two values, so a multiplexer with a one-bit selector can choose between two inputs. This device will have two one-bit inputs, one one-bit selector, and one one-bit output. A quick table can summarize the exact behavior.

Input0	Input1	Selector	Output
0	0	0	0
1	0	0	1
0	1	0	0
1	1	0	1
0	0	1	0
1	0	1	0
0	1	1	1
1	1	1	1

This table is much larger than previous ones, but is still fairly simple. Whenever the selector is zero, the value at the first input is placed on the output line. Whenever the selector is one, the value at the second input is sent to the output.

The multiplexer behavior table illustrates one of the many examples in which it is convenient to use numbers starting with zero. Input0 is selected when the selector contains the value zero, and Input1 is selected when the selector contains the value one.

As usual, examining the behavior table reveals how this device can be implemented. Conceptually, the multiplexer "forwards" the correct input signal to the output. To implement the device, we only need to concern ourselves with when the output is one and when it is zero.

In this case, the output is one when:

- Input0 is one AND Selector is NOT one

  OR

- Input1 is one AND Selector is one

This can be directly converted to a gate-level schematic:

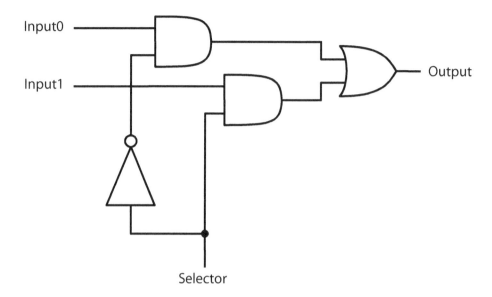

The selector sends a one to the bottom input on the appropriate AND gate. With one of its two inputs held at one by the selector, the selected AND gate will output a one whenever its input line is a one, and zero whenever its input line is zero. The unselected AND gate will have one of its inputs set to a zero by the selector, causing it to never output a one.

This is a simple one-bit multiplexer, and can be represented with its own symbol.

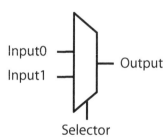

Extending this design to accommodate multi-bit data is quite straightforward. The one-bit multiplexer above can choose which input a bit should be read from. To create two-bit multiplexer we can simply use two one-bit multiplexers being issued the same selector signal. The first multiplexer will choose which input the first bit should come from, and the second multiplexer will choose which input the second bit should come from.

In schematic form:

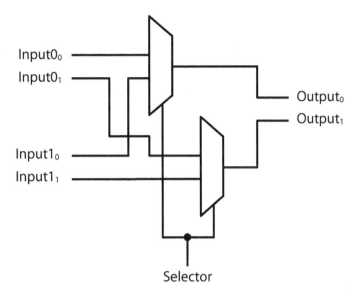

The first two-bit input – Input0 – enters at the top. The second two-bit input – Input1 – enters at the bottom. When the selector has a value of zero, both multiplexers output their top input – the two bits that make up Input0. When the selector has a value of one, both multiplexers output their bottom input – the two bits from Input1.

This multi-bit pattern can be expanded to as many bits as necessary. Just as two multiplexers can choose between two two-bit numbers, ten multiplexers could choose between two ten-bit numbers. These multi-bit multiplexers can be represented with a more concise symbol:

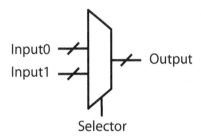

The slashed lines indicate that the inputs and outputs consist of multiple wires (or bits). As we are still only selecting between two (multi-bit) inputs, the selector only needs to be a one-bit number, so it is only a single wire. This brings us to the next logical step: more than two inputs to choose from.

Thankfully, this extension can also be implemented using multiple multiplexers. The last multiplexer we looked at could choose between two (multi-bit) inputs. We can re-use these multiplexers several times to create a hierarchy of choices. If we had four inputs, we would need two two-input multiplexers to accommodate all of the inputs. We would then have two outputs, which would require one more two-input multiplexer to select the appropriate output.

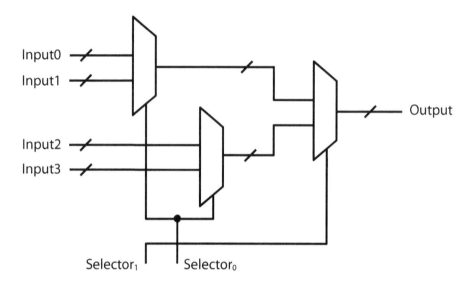

The inputs and outputs still carry multiple bits, as before. The selector is now also a multi-bit signal. When the selector carries the value $(0)_{10}$ / $(00)_2$ the left-hand multiplexers output their top inputs (Input0 and Input2), and the right-hand multiplexer outputs its top input (Input0). When the selector has a value of $(3)_{10}$ / $(11)_2$ the left-hand multiplexers output their bottom inputs (Input1 and Input3), and the right-hand multiplexer outputs its bottom input (Input3).

Each bit of the selector, starting from the right, halves the number of choices. In the four-input example, when the rightmost bit is set to one, the left two multiplexers narrow the choice to either Input1 or Input3. This correlates exactly to the binary number being sent on the selector wires; if you know a two-bit binary number has its rightmost bit set to one, you can infer that it is either a $(1)_{10}$ or a $(3)_{10}$.

Taking this a step further, you can imagine how an eight-input multiplexer could be created. The selector would have to grow to three bits to accept eight different values.

Four two-input multiplexers would be required to accept all eight inputs. The rightmost bit of the three-bit selector would narrow the eight choices down to four. If the rightmost bit was a one, the four leftmost multiplexers would output Input1, Input3, Input5, and Input7 – the only possible three-bit numbers with a rightmost bit of one. These four inputs would be fed into two more two-input multiplexers, whose output would be fed into one final two-input multiplexer.

One more column of multiplexers would create the following sixteen-to-one multi-bit multiplexer.

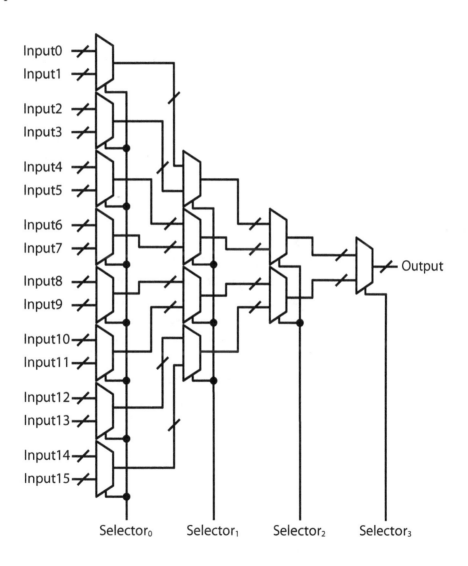

Regardless of how many inputs or bits we have, we can draw a simplified symbol to represent a multi-bit, multi-input multiplexer.

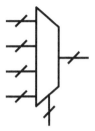

You can see that this multiplexer accepts four multi-bit inputs and produces one multi-bit output. From this, you can infer that the multi-bit selector at the bottom must consist of two bits (to accommodate the selector values zero through three). From this picture alone it is not clear how many bits are in each input and the output, but it may be labeled beside the slashed input or output lines. If the inputs (and output) were five bits wide, you could refer to this device as a "five-bit four-to-one multiplexer", as it converts four inputs to one output, each of which is five bits in size.

5.5 Demultiplexing

The next simple machine is the opposite of the multiplexer: the demultiplexer. This device accepts one input, and forwards it to one of several outputs, as indicated by the value on the selector wires. The simplest case is, of course, one input, two outputs, and a one-bit selector. The specific behavior is summarized in the following table:

Input	Selector	Output0	Output1
0	0	0	0
1	0	1	0
0	1	0	0
1	1	0	1

Once again, we can translate this table to words.

Output0 is one when: Input is one AND Selector is NOT one.

Output1 is one when: Input is one AND Selector is one.

Converting these sentences into gates is also fairly straightforward:

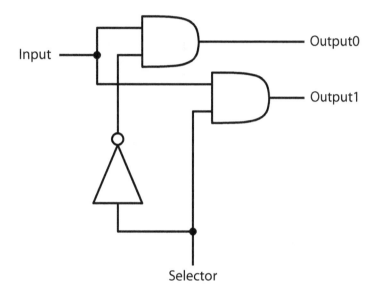

The demultiplexer symbol is very similar to the multiplexer symbol.

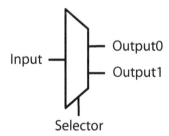

Just as was done with multiplexers, multiple single-bit demultiplexers can be used to create a multi-bit demultiplexer.

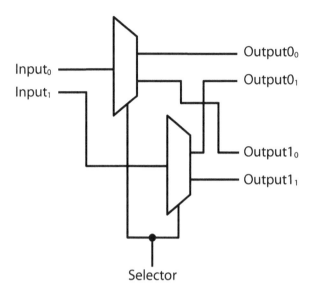

The two-bit one-to-two demultiplexer shown above can be represented with the following symbol:

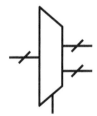

As one final extension, we can also use several multi-bit two-output demultiplexers to create multi-bit demultiplexers with more than two outputs.

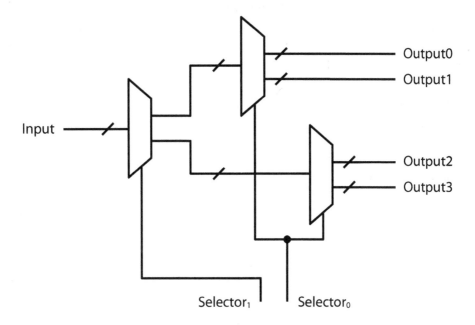

In general, a multi-bit multi-output demultiplexer can be drawn using a symbol similar to the following.

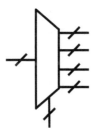

This particular demultiplexer has four outputs, so the selector must contain two bits. This diagram does not specify the number of bits in the input and outputs.

5.6 Even More Memory

We now have devices that can store a number (registers), and devices to select from a variety of inputs and outputs (multiplexers and demultiplexers). By combining these devices, we can create a new device that can store many numbers, with the ability to select which one we would like to retrieve (read) or set (write). In particular, we need a device that can store numbers, change any of the numbers one at a time, and retrieve two numbers at a time. We will refer to this device as a "register block".

We will need three inputs to write to a selected register: the address (or number) of the register to write, the value to write into the register, and a signal to indicate whether or not writing is enabled.

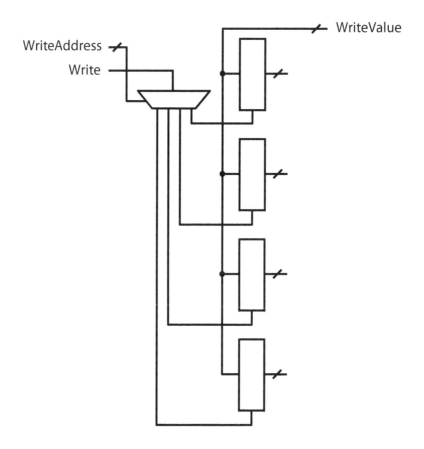

The value to be written is delivered to the inputs of all registers, but only one register should be changed. To accomplish this, a demultiplexer uses the write address to determine where to deliver the "write" signal. Recall that a demultiplexer simply connects its input to one of its outputs. To write to a register, it will not only have to be selected by the demultiplexer, but the "write" signal must also be set to true.

The schematic so far solves two of the requirements. This device can store several numbers, as well as change (write to) one of the locations at a time. The final requirement is the ability to read two of the values at a time. All four registers are outputting the values they contain; we simply need to select which ones to connect to the two output connections. This is a job for a multiplexer (or two).

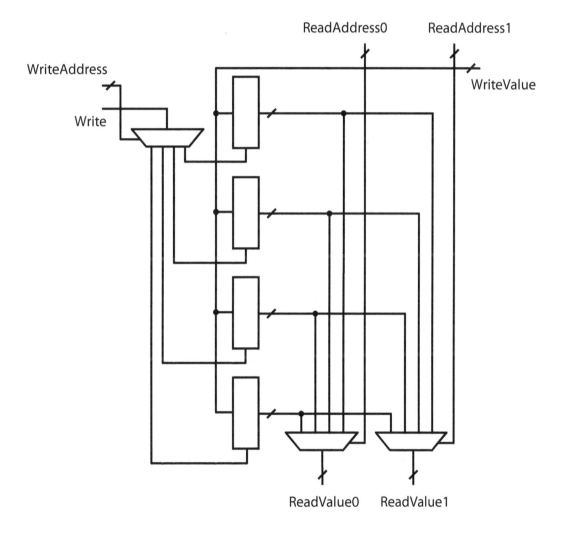

Although the schematic became significantly larger, the changes are relatively simple. All four outputs are connected to each of the multiplexers. Each multiplexer also has an address connection to select which of the four inputs (coming from the four register outputs) to connect to their output.

The simplified symbol for this device is as follows.

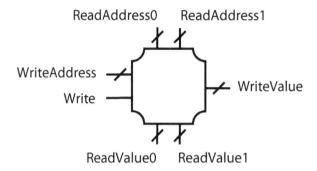

The bottom two lines are outputs, while the rest are inputs. The two read-value connections output the contents of the registers selected by the two read addresses. Values are stored by selecting where to store the value via "WriteAddress", inputting the value to store via "WriteValue", and setting the "Write" line to true to carry out the write.

The sizes of the connections are not labeled. The width of the connections is determined by two factors: the size of the stored values and how many of them can be stored. The value connections are all the same size as the stored data. If the device stores sixteen-bit numbers, each of the "WriteValue", "ReadValue0", and "ReadValue1" connections must have sixteen wires to accommodate the numbers that can be read and written. The size of the address connections is determined by how many values the device can store. If the device can store thirty-two values, each of the address connections must contain five wires to be able to uniquely identify each of the thirty two locations (zero through thirty one).

5.7 Arithmetic

Reading and writing values is certainly helpful, but to make a computing device even more useful, it needs the ability to perform some calculations. A simple adding unit can provide a great deal of functionality thanks to the two's complement numbering system. The chapter on binary revealed that adding two's complement numbers works regardless of whether inputs are positive or negative. As a result, a single adder can add both positive and negative numbers.

Recall that adding two multi-bit binary numbers is actually just the repeated action of adding the columns of single-bit digits. The result of adding a column of two single-bit numbers is summarized in the following table. Input0 and Input1 come from the numbers being added. Adding them together produces a sum (and possibly a carry).

Input0	Input1	Sum	Carry
0	0	0	0
1	0	1	0
0	1	1	0
1	1	0	1

Converting this behavior to gates is simple. The sum is one whenever exactly one of the inputs is true, and a one is carried whenever both of the inputs are true.

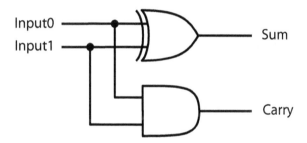

This adder would be sufficient for adding single-bit numbers, but we need to do more. In addition to generating a carry output, the adder needs to accept a carry input to be a part of multi-bit addition. Recall that to add binary numbers, each column is added, with a one sometimes carried to the next column to the left. We will call the action of sending a one to the next column a "carry out". When the column to the left is added, the one that was carried over must be taken into account; we can refer to this as a "carry in". This adder needs to be expanded to create the following behavior.

CarryIn	Input0	Input1	Sum	CarryOut
0	0	0	0	0
0	0	1	1	0
0	1	0	1	0
0	1	1	0	1
1	0	0	1	0
1	0	1	0	1
1	1	0	0	1
1	1	1	1	1

Before converting this large table to gates, we can convert it to a set of sentences that describe the behavior.

The sum in this new adder is one when:

- Exactly one of the Inputs is one

  XOR (exactly one of these two is true)

- CarryIn is one

CarryOut in this new adder is one when:

- Exactly one of the Inputs is one AND CarryIn is one

  OR

- Input0 is one AND Input1 is one

Translating these two descriptions into gates results in the following:

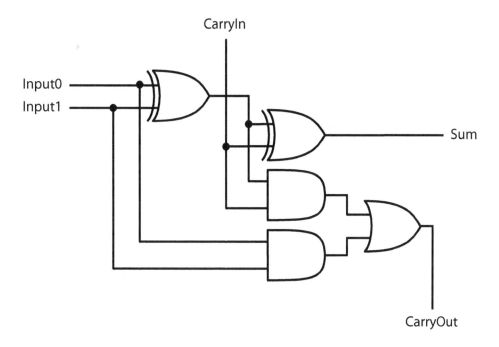

This device is commonly referred to as a "full adder". It is capable of adding a single column of a multi-bit addition, taking into account the values that may be carried in and out.

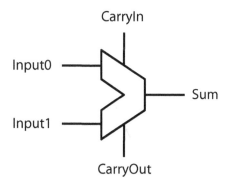

Adding a multi-bit number can now be accomplished by chaining together the devices capable of adding each column. The addition of two four-bit numbers has four columns, requiring four full adders.

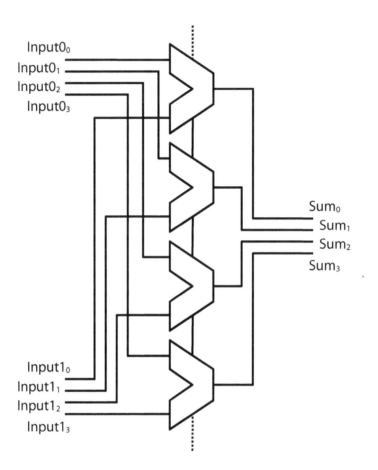

As in the binary adding examples, any final carry out is discarded to make the result fit into the correct number of bits (four, in this case). Each bit of the first input is directed to the first input of a column adder, and each bit of the second input is directed to the second input of a column adder. Each column adder will add the two input bits and send the sum to the corresponding output bit, as well as send the carry-out bit to the next adder's carry-in connection.

When the value on an adder's carry-in connection changes, the adder's sum output and carry-out output may change as well. This can cause a chain reaction, as each adder after the first can eventually receive a carry-in value when the previous adder finishes adding. The name "ripple carry adder" is often used to describe this device, as the carry from one column ripples to the next, until all bits have settled to their final values.

In general, a multi-bit adder is drawn as follows.

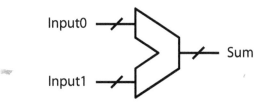

5.8 Exercises

1. Consider the following two sequences when using a four-bit register. What value does the register store at the end of each one? What value appears on the register's output wires at the end of each one?

    a. Set the input to 3. Set the write line to true. Set the input to 5. Set the write line to false.

    b. Set the input to 3. Set the input to 5. Set the write line to true. Set the write line to false.

2. How many two-input multiplexers are required to select from sixty-four inputs?

6

Multi-Purpose Machines

6.1 Rearranging Machines

The previous chapters illustrated the design of several simple machines. Each machine could perform a simple task, be it switching, adding, or storing data. Just as gates were interconnected to create simple machines, these simple machines can be connected to create more complex machines.

Multi-step processes can be performed by changing the connections between several simple machines in some pre-determined order. As a simple example, consider using the register block and adder that we designed earlier. With the appropriate series of interconnections, these devices can add four numbers from four incoming connections. To be specific, we will use a register block with sixteen registers, each of which holds a sixteen-byte number. Because this register block contains sixteen registers, we need four bits to specify a register number (address). This means that the write address and each read address is a four-bit (or four-wire) connection. Because each register holds a sixteen bit value, the write-value input and each read-value output consist of sixteen wires. Correspondingly, we will use a sixteen-bit adder, accepting two sixteen-bit inputs and producing a sixteen-bit sum.

In the following diagrams, the four input connections enter at the top-left, and the result output exits at the bottom.

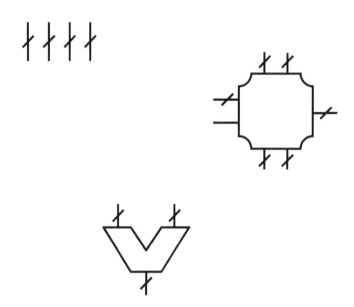

Because the adder can only add two numbers at once, we could do the following:

1. Add the first two numbers, storing the result in location zero of the register block.
2. Add the second two numbers, storing the result in location one of the register block.
3. Add the two numbers in location zero and location one of the register block.

At the end of these three steps, the adder's output would contain a value representing the sum of all four numbers. Taking a closer look at this process, you can imagine how the connections would have to change at each step. The following diagrams will make the necessary wire connections, as well as specify the appropriate values on remaining register block inputs.

1. Add the first two numbers, storing the result in location zero of the register block.

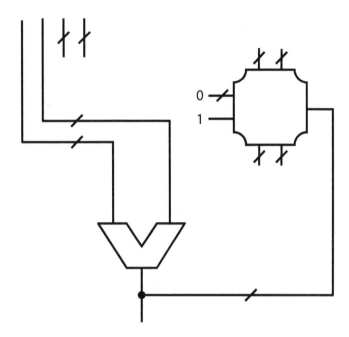

2. Add the second two numbers, storing the result in location one of the register block.

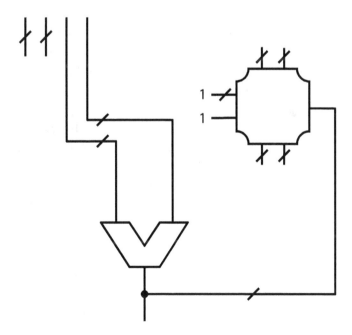

3. Add the two numbers in location zero and location one of the register block.

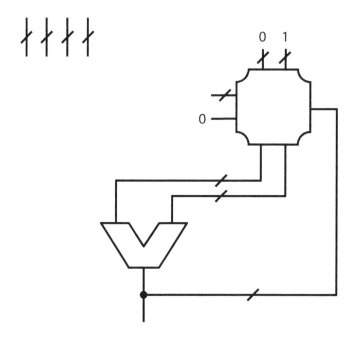

Following the steps above would save you the effort of manually adding all four numbers. As you may have guessed, physically rearranging the connections between each step would be quite tedious and error-prone. Thankfully, we have already designed a device that can switch between various connections: the multiplexer.

Instead of constantly rearranging the wires between simple machines, we can create every connection between the devices that we will ever need, and then use multiplexers to select the appropriate connection for a specific step. In the adding example above, we could put a multiplexer in front of each of the adder's inputs to choose between inputting from the number inputs (needed in steps one and two), or the output coming from the register block (needed in step three). Rather than reconnecting physical wires, we only need to set the appropriate values on the selector inputs of the multiplexers, and the multiplexers will create the appropriate connections for us.

The two multiplexers could be connected in the following manner:

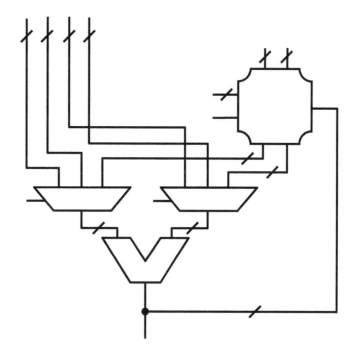

Step one would require a selector value of zero be sent to both multiplexers, to select the first two input numbers. Step two would require a value of one to be sent, to select the second pair on input numbers. Finally, step three would require a value of two, to select the values from the register block.

6.2 Instructions

While you could physically disconnect a wire and re-connect it anywhere you please, multiplexers can only choose from the signals already connected to their inputs. To use multiplexers to make all of the connection changes, we need to connect all possible inputs to the multiplexers ahead of time. To determine which connections are necessary, we need to decide on which arrangements of simple machines we want to be possible. We will use arrangements that can accomplish the following:

Set Register: This arrangement will save a specified number into a specified location in the register block.

Copy Register: This arrangement will copy the value in a specified register into another specified register.

Add: This arrangement will set a specified register's value to the sum of the numbers contained in two other specified registers. The destination register cannot be one of the two source registers.

Print: This arrangement will send the contents of a specified register to some external wires, so an external device can see the value. It will also set an external "output signal" connection to high voltage so the external device knows a value has been sent.

Because we are using multiplexers to make the appropriate connections, obtaining each arrangement only requires the correct values be set on the selector inputs of the multiplexers. Consider the "Add" arrangement, in which a sum is to be saved to a register. Multiplexers would have to be given whatever selector values are necessary to cause the output of the adder to be connected to the write-value input of the register block.

You may have noticed that each arrangement has one or more "specified" values. To actually carry out the operation, we need to know not only which values should be set on the multiplexer selectors, but also these "specified" values. For example, once the multiplexers have made the proper arrangement of simple machines for the "Set Register" operation, the device still needs to know which register to write to, and what number should be written.

To carry out one of the steps in a process, the device needs to know which operation to carry out (Add, Print, etc.), which indicates how the simple machines should be interconnected, as well as any data that needs to be specified (which number to store, which registers to add, etc.). Each of the "specified" values is referred to as an "option" or "argument". The combination of operation and arguments is known as an "instruction". Using the four operations specified above, we can create a set of instructions to add four numbers, and then print the sum.

Note that there is no operation that can add two specified numbers. The add operation can only add the contents of registers in the register block. To add numbers, the numbers will first have to be stored into the register block using the "Set Register" operation. For the sake of example, consider adding the numbers four, five, six, and seven. We will use an adder and a register block that contains sixteen registers:

1. Set Register – write four into register number zero of the register block.
2. Set Register – write five into register number one of the register block.
3. Set Register – write six into register number two of the register block.
4. Set Register – write seven into register number three of the register block.
5. Add – add the contents of register number zero and register number one, and store the result in register number four.
6. Add – add the contents of register number two and register number three, and store the result in register number five.
7. Add – add the contents of register number four and register number five, and store the result in register number six.
8. Print – output the contents of register number six.

The eight instructions listed above can be carried out by a machine; they specify which operation to perform (indicating how the simple machines should be interconnected), as well as specifying the values used in each operation. A sequence of instructions that a machine can carry out is known as a "program".

One final operation is useful for creating more complicated programs. The program listed above simply performs each instruction in order. To gain a bit more flexibility, we can create an instruction that indicates that the machine should "jump" to a different location in the list of instructions.

> **Loop**: This arrangement will cause the device to carry on executing at a specified position in the list of instructions, if the value in a specified register is not zero.

Loop is the final instruction our multi-purpose machine will understand. The list of all possible instructions that a machine can understand is referred to as its "instruction set". The instruction set we have created contains five instructions.

6.3 Instruction Encoding

While a machine can carry out the instructions to add four numbers, it cannot read them in the form they have been written. The simple machines, such as the adder and register block, can't accept descriptive sentences as input; they can only accept the high and low voltage signals that we use to represent binary numbers. To convey these instructions to the machine, we will have to devise a way to represent them in voltages (or the binary that the voltages represent).

For the sake of simplicity, every instruction will be the same size. Each instruction must contain the operation to carry out, as well as several arguments. The maximum number of arguments occurs in the "Add" operation, which requires three arguments: two register numbers from which to read and the number of a register in which to write. As our example has sixteen registers in the register block, we need four bits to represent a register number (zero through fifteen). This choice of size is fairly convenient, as four bits maps to exactly one hexadecimal digit. To continue this trend, we can make the operation part of the instruction a four-bit number as well. This allows us to express a full instruction as a four-digit hexadecimal number. The first digit will be the operation, with the following three digits containing the arguments. Consider the following number (shown in both binary and hexadecimal):

$$(\ 1000 \ 0101 \ 1010 \ 1110 \ )_2 \qquad (\ 85AE \ )_{16}$$

When interpreted as an instruction, this value would represent operation number eight, with arguments five, ten, and fourteen.

Which four-bit numbers we use to represent each type of operation is fairly arbitrary. We will use numbers that are easy to identify at a glance.

Operation	Binary	Hexadecimal
Print	0000	0
Set Register	0001	1
Copy Register	0010	2
Add	0100	4
Loop	1000	8

Each instruction also has three spaces for arguments. We must decide in which of the three spaces each argument should be placed.

Operation	Argument0	Argument1	Argument2
Print		Number of the register to output	
Set Register	Number of the register to write to		Number to write
Copy Register	Number of the register to copy to	Number of the register to copy from	
Add	Number of the register to write to	Number of the first register to add from	Number of the second register to add from
Loop		Number of the register to check for zero	Number of the instruction to jump to

Using these two tables, we can translate the eight-instruction program from the previous section into instructions that a machine can understand.

1. Set Register – write four into register number zero of the register block.
2. Set Register – write five into register number one of the register block.
3. Set Register – write six into register number two of the register block.
4. Set Register – write seven into register number three of the register block.
5. Add – add the contents of register number zero and register number one, and store the result in register number four.
6. Add – add the contents of register number two and register number three, and store the result in register number five.
7. Add – add the contents of register number four and register number five, and store the result in register number six.
8. Print – output the contents of register number six.

In hexadecimal (remember, the first digit is the operation, followed by three arguments):

```
1004
1105
1206
1307
4401
4523
4645
0060
```

In binary:

```
0001 0000 0000 0100
0001 0001 0000 0101
0001 0010 0000 0110
0001 0011 0000 0111
0100 0100 0000 0001
0100 0101 0010 0011
0100 0110 0100 0101
0000 0000 0110 0000
```

The instructions are now encoded into a format that the machine can understand. These numeric instructions are referred to as "machine code" or "byte code".

This is not the only possible translation of these instructions. Some instructions do not use all three argument spaces. For example, the final print instruction only needs bits twelve through fifteen set to zero (the print operation), and bits four through seven set to six (the register number to print). The other bits (which contain argument zero and argument two) can be set to anything; the print instruction does not use these arguments. The machine code 0x0F6F is an equally valid translation of the final instruction.

An emulator for a CPU capable of executing these instructions has been provided in the accompanying resources. In the folder for this section – 06.03 – clicking "run.bat" will run the emulation. The emulator will execute the machine code contained in the instructions.txt file. Each line of instructions.txt is a machine code instruction in hexadecimal format. Anything after a semicolon is simply a comment, and is ignored by the emulator. Feel free to experiment with different programs by editing instructions.txt, then clicking "run.bat" again.

6.4 Sign Extension

The instruction encoding we have chosen leads to an interesting situation: each register in the register block consists of sixteen-bits of storage, while the "set register" instruction can only specify a four-bit number to store. We could simply set the lowest (rightmost) four bits of the register – bits zero through three – to the four-bit number in the set-register instruction and set bits four through fifteen to zero. This would mean we could use the set-register instruction to set values zero through seven (setting 0x0 through 0xF would store 0x0000 store 0x000F). Although this is a completely valid solution, a different solution is also possible.

Rather than copying in the four specified bits and setting the rest to zero, we can copy in the four specified bits and set each of the remaining bits to the same value as the leftmost bit of the four-bit number. This means that a set-value instruction with a four-bit argument of $(1010)_2$ would actually set the register's sixteen-bit value to $(1111\ 1111\ 1111\ 1010)_2$. A set-value instruction with an argument of $(0101)_2$ would still set the register's value to $(0000\ 0000\ 0000\ 0101)_2$. This technique is referred to as "sign extension".

Recall that there are two interpretations of the four-bit value $(1110)_2$ (0xE in hexadecimal). If we are treating numbers as unsigned (always positive), $(1110)_2$ is fourteen. If we are treating numbers as signed (possibly negative), then $(1110)_2$ is negative two (in two's complement). The choice of how to expand a four-bit number into a sixteen-bit space depends on whether we would like to use signed or unsigned numbers.

We would want to use the first implementation – setting any extra bits to zero – if we are using unsigned numbers. When using unsigned numbers, the value $(1110)_2$ represents fourteen. Thus, when we perform a set-register instruction with a value of $(1110)_2$, we would like the register to store the value $(0000\ 0000\ 0000\ 1110)_2$, because that is the sixteen-bit representation of fourteen.

If using signed numbers, $(1110)_2$ represents negative two. In this case, we would want the register to be set to the sixteen-bit representation of negative two: $(1111\ 1111\ 1111\ 1110)_2$. This is exactly what sign extension would perform. Because we will use the ability to set a register to a negative value, we use the sign-extension method of expanding numbers.

Whether using sign extension or not, the four wires with the instruction argument will be connected to the write-value input of the register block so that the register block can store the four specified bits. The difference is how the remaining wires are connected. In the unsigned case, the write-value wires number four through fifteen would be connected to zero (low voltage). In the signed case (using sign extension), wires number four through fifteen would all be connected to wire number three, so all of the remaining bits receive the same value as the leftmost bit of the specified four-bit value.

6.5 Assembly

Although we decided to design machines based on binary numbers, we also examined the hexadecimal numbering system. Binary numbers are tedious for people to manipulate, and decimal numbers aren't very closely related to their binary equivalents. Hexadecimal is a good compromise, as it is very easily related to binary (four bits = one hexadecimal digit), but more convenient for us to write.

The same type of situation can be seen with computer instructions. The descriptions of instructions – "Print", "Set Register", etc. – are not closely related to the machine code, but machine code is quite tedious for people to deal with. A good compromise for writing instructions that can be easily translated into machine code is a language called "assembly". The assembly language uses short statements similar to natural language that much more closely represent the machine code.

The specific words in the language vary from one instruction set to the next. We will use the following abbreviations:

Operation	Abbreviation
Print	print
Set Register	set
Copy Register	copy
Add	add
Loop	loop

In the assembly language, instructions are written as an operation followed by the arguments that it requires. For our instruction set, we will write arguments in order (Argument 0, Argument 1, and then Argument 2). We will simply omit any arguments that don't apply to an operation (e.g. Argument0 has no purpose in the print operation). Numbers that specify a value will be written in their normal form, while numbers that specify a register will be prefixed by "#". When converted to machine code, the hash mark is simply ignored. Anything that appears after a semicolon (on the same line as the semicolon) is also ignored. This allows us to write notes or "comments" in the code to better explain what it does. These comments are solely for our understanding; they have no effect on the behavior of the instructions.

Writing the previous eight-instruction program in assembly can be done as follows:

```
set #0, 0x4        ; store four into register 0
set #1, 0x5        ; store five into register 1
set #2, 0x6        ; store six store six into register 2
set #3, 0x7        ; store seven into register 3
add #5, #0, #1     ; add register 0 and register 1 into register 5
add #6, #2, #3     ; add register 2 and register 3 into register 6
print #6           ; display the result
```

With a more convenient way to write instructions, we can now create a more complicated program with less effort. We will compose a program that can calculate the twenty-fifth Fibonacci number. The Fibonacci number sequence is a list of numbers that follows a simple pattern: the first number is zero, the second number is one, and each subsequent number is the sum of the previous two.

$$0, \ 1, \ 1, \ 2, \ 3, \ 5, \ 8, \ 13, \ 21, \ 34, \ 55, \ 89, \ ...$$

The first few numbers are easy to compute manually, but they grow very quickly as the sequence progresses. The twenty-fifth number is over forty-five thousand. This is a good candidate for a task that should be automated.

We could manually write the first twenty-four Fibonacci numbers on paper by carrying out the following steps:

1. The first number is zero, write a zero.
2. The second number is one, write a one.
3. Add the last two numbers and write the result; this is now the last number. What used to be the last number is now the second-last number.
4. Check if step 3 has been done twenty-three times. If not, go back to step 3.

If you carried out these steps, you would eventually have the twenty-fifth number of the Fibonacci sequence. Take note that step three is only repeated twenty-three times because the first two numbers are already known before reaching step three. Because the first two numbers are simply given, not calculated, our program will just display the last twenty-three.

We can start by storing the two starting numbers. We can use register one to store the last number, and register zero to store the second-last number.

```
set #0, 0x0        ; set the second-last number to 0
set #1, 0x1        ; set the last number to 1
```

We now need to repeatedly add the last two numbers. There is only one instruction that allows steps to be repeated – the loop instruction. Loop will "jump" to a specified step if a specified register is not zero. Therefore, to repeat a step twenty-three times, we can start by storing the number twenty-three in a register to be used as a counter. We then decrease the value in that counter register each time we carry out the step, followed by loop instruction. The loop instruction will jump back to the specified step until the value in the counter register reaches zero, at which point the loop instruction will do nothing.

You may have noticed a slight problem; we don't actually have any instructions capable of storing the number twenty-three in a register. The spaces for each argument of an instruction are only four bits wide. Thankfully, some simple addition can solve this problem.

```
set #0, 0x0         ; set the second-last number to 0
set #1, 0x1         ; set the last number to 1
set #2, 0xF         ; set register 2 to -1, remember sign extension
set #4, 0x6         ; set register 4 to 6
add #3, #4, #4      ; set register 3 to 12 (6 + 6 = 12)
add #4, #3, #3      ; set register 4 to 24 (12 + 12 = 24)
add #3, #4, #2      ; set register 3 to 23 (24 + -1 = 23)
```

The value twenty-three is now in a register to be used as a counter. We can now repeatedly calculate the Fibonacci numbers. Note that instruction numbers start at zero.

```
set #0, 0x0         ; set the second-last number to 0
set #1, 0x1         ; set the last number to 1
set #2, 0xF         ; set register 2 to -1, remember sign extension
set #4, 0x6         ; set register 4 to 6
add #3, #4, #4      ; set register 3 to 12 (6 + 6 = 12)
add #4, #3, #3      ; set register 4 to 24 (12 + 12 = 24)
add #3, #4, #2      ; set register 3 to 23 (24 + -1 = 23)
add #4, #0, #1      ; add the last number and second-last number
print #4            ; display the calculated Fibonacci number
copy #0, #1         ; the last number is now be the second-last number
copy #1, #4         ; the sum we just calculated is now the last number
add #4, #3, #2      ; decrease the counter value
copy #3, #4         ; store the decreased value in the counter register
loop #3, 0x7        ; if the counter isn't 0 yet, jump to instruction 7
```

The Fibonacci program is now complete. The program stores the two initial values (zero and one), and then initializes a counter (register number three) to twenty-three. The program then calculates the next Fibonacci number twenty-three times, outputting the new value and updating the "last" and "second last" value registers each time.

The additional resources bundle contains a machine code version of this program that can be used with the emulator. Clicking run.bat in the folder for this section – 06.05 – will run this program in the emulator.

6.6 Exercises

1. The machine-code version of the Fibonacci program in the additional resources bundle has two extra instructions at the end (0x1001 and 0x800E).

    a) What do these instructions do?

    b) Why are they here?

2. Extend the following program to multiply the positive, non-zero values in register zero and register one, then print the result. Recall the restrictions on the "add" operation (see section 6.2).

```
set #0, 5              ; store the first number to be multiplied
set #1, 6              ; store the second number to be multiplied
add more instructions here
print #2               ; output the result
set #1, 1              ; make sure the loop is taken
loop #1, 0             ; do it again, just for fun
```

3. Convert your program from the previous question into machine code. Enter it in instructions.txt in the folder for this section (06.06) and see if it works. Try changing the values in the first two instructions to multiply different values (recall that any number larger than 0x7 is treated as negative and will be sign-extended).

7

The CPU

7.1 Terminology

With a fully specified instruction set, and some sample programs, we can now design some hardware that can make use of these instructions. Because these devices are so generic, carrying out whatever instructions are issued to them, they are often the heart of various computing devices. This central role led to the common name "central processing unit", or "CPU".

A CPU is the device that carries out each instruction. To cause the device to stop carrying out one instruction and start performing the next instruction, a periodic signal must be issued to the CPU. This signal continuously cycles between high voltage and low voltage, and is called the "clock". In our CPU, each full cycle of the clock (a period of time with the signal at high voltage, followed by a period of time with the signal at low voltage) completes one instruction. The switch from low voltage to high voltage causes the switch from the current instruction to the next instruction.

Because the clock causes the CPU to move to the next instruction, the clock signal cycle should not be made any quicker than the settling time required to complete an instruction. This clock speed is one of the primary measures of a CPU's performance, and is measured in a multiple of hertz (megahertz, gigahertz, etc.). This measure describes the number of times per second that the CPU clock completes a full cycle from high voltage to low voltage and back to high voltage. Because the CPU we are designing completes one instruction per cycle, a clock speed of ten megahertz would imply that ten million instructions can be completed per second.

7.2 Instruction Decoding

The versatility of a CPU comes from its ability to dynamically change how it behaves by altering how its components are interconnected. The instruction specifies how the CPU should behave in the four bits that hold the operation. For example, we decided that the "add" operation should add two values from the register block, and store the sum in the register block. Thus, when the "add" operation is issued, the CPU should connect read-value outputs of the register block to the inputs of the adder. The CPU should also connect the sum output of the adder to the write-value input of the register block.

The instructions that arrive at the CPU will specify which operation to perform, not how to interconnect the various components. This allows the same instructions to be used on CPUs that have different internal structures. Unfortunately, this also means that the operation identifier must be converted to a set of signals that properly change how the CPU behaves. The part of the CPU that performs this conversion is called the "controller".

The controller simply controls the other parts of the CPU, so it can't be implemented until we know what other parts we are using. For now, the controller will simply be drawn as a blank square. Any part of the CPU that needs to change its behavior based on the type of operation being executed will be connected to an output of controller. Anything that indicates how the CPU should behave will be an input to the controller.

Although the instructions for this device are sixteen-bit numbers, the entire sixteen-bit number itself isn't very useful; each instruction contains four individual segments. The first step in processing the instruction is to break apart the sixteen-bit signal into the four individual four-bit signals that it contains. We can accomplish this by simply regrouping the sixteen wires that carry the instruction. The four bundles of wires carry the operation, and arguments zero through three.

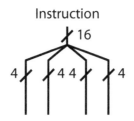

Bits twelve through fifteen (the leftmost four bits) specify which operation to perform. This information needs to be sent to the controller, so we can connect these four wires to the controller's input.

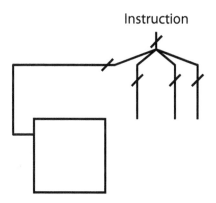

Instruction

7.3 Registers

Every instruction refers to either writing to or reading from some registers. To carry out these operations, we will need a register block. You can see that whenever an instruction writes to a register (set, copy, and add) the number of the register to write to is stored in the four bits next to the operation (bits eight through eleven). This implies that the write-address input of the register block should be connected to these four bits. When the set, copy, or add operations are issued, the four bits that specify which address to write to will be sent to the write-address input of the register block.

You can also see that when an instruction reads from a single register (print, copy, and loop), the number of the register to read from is in bits four through seven. We will connect the first read-address input to these four bits. This means that whenever an instruction reads from a single register, the value that is read will appear on the first read-value output of the register block. When an instruction also specifies a second register to read from (add does this), the second register number is in bits zero through three. This means that we should connect the rightmost four instruction wires to the second read-value input of the register block.

Making all three of these connections yields the following:

Instruction

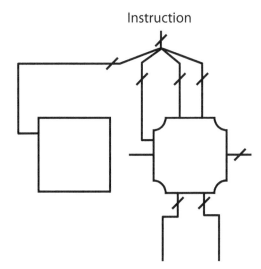

You can imagine placing an instruction on the top connection and follow where each section of the instruction will be delivered. The operation will be sent to the controller, to decide how the other components should function. Argument zero will be sent to the write-address input of the register block, and arguments one and two will be sent to the two read-address inputs.

Note that because these wires are always connected, whatever value is in each part of the instruction will be sent to the corresponding location shown above. For example, the print operation does not write to a register, but bits eight through eleven of the instruction, whatever they may be, will still be sent to the write-address input of the register block. It is up to the controller to make sure that the CPU behaves correctly no matter what value is in that space. In other words, the write-address section of the instruction should be effectively ignored for the print operation. The controller must ensure it actually is ignored.

7.4 Timing

We have decided where the register block will obtain the write-address input, but haven't discussed what will enable or disable writing via the "write" connection. When the write connection is set to the high voltage, whatever value is set on the write-value input will be written to the register identified by the write-address input. In short, whenever the write connection is set to high voltage, something inside the register block will be overwritten. This means that it is very important that the write line only be set to true when a value actually should be written. Whether or not a register should be written is determined by which operation is being executed, meaning the write line should be connected to the controller.

In addition to only the setting write connection to high voltage when it is required by an operation, it is also important that the correct write address has reached the register block before the write line is set to true. If the "write" connection reaches the register block before the "write address" does, a write could take place at an undesired location. Although we can't make the write-address value move faster, we can make the write-enable value move more slowly.

The time it takes a voltage change to propagate depends mostly on how many transistors must switch between the change and its destination. To make a signal propagate more slowly, we can simply insert additional gates in its way. The simplest choice is a string of NOT gates. So long as an even number of NOT gates are used, the output value will be the same as the input value, only slightly delayed. We will represent a delay component with the following symbol:

To delay the write connection, we can make it depend not only on the current operation, but also on a delayed clock signal. In particular, the write connection will be set to high voltage if the operation requires a write AND the delayed clock is high voltage. Recall that the clock signal changing to high voltage is what triggers the next instruction to begin being processed. Soon after the clock signal goes to high voltage, the new write address will reach the register block. By making the write-enable connection depend on the clock being high, and delaying the clock signal to the control block (which generates the write-enable signal), we can be sure that the write-enable connection will only change to high voltage after the new write address has arrived at the register block. The write connection will only be set to true if the operation indicates a write should take place, and a short delay has elapsed since the instruction began.

The following diagram includes a delayed clock signal being sent to the control unit. The control unit can now use the delayed clock signal and the operation value to determine whether or not to set the register-write connection to high voltage.

Clock   Instruction

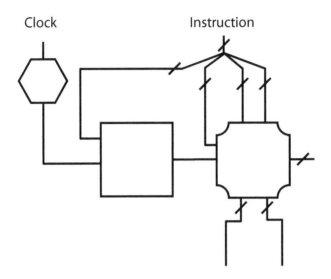

7.5 Storing Results

We have now connected everything we need to decide when to write values and where to write them. The next step is to make the connections required to determine what value to write. The register read-addresses and write-address are always specified in the instruction, so they are simply connected with wires. Unlike these connections, the value to write can come from several different places. This implies that the write-value connection should be connected to a multiplexer, so the controller can choose from where the write value should be obtained.

The controller chooses the source of the write value based on the instruction, so the selector input of the multiplexer is connected to the controller.

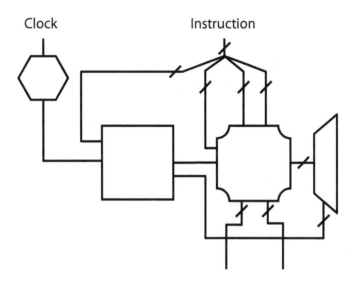

We can now connect all the possible sources of write values to the multiplexer. The "set" operation stores a value from bits zero through three of the instruction. This value can be sent to the multiplexer by simply connecting to the four rightmost wires of the instruction.

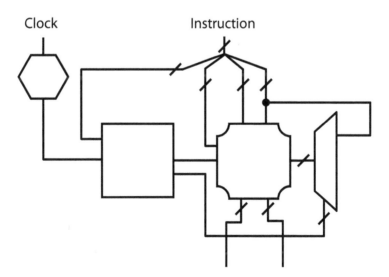

In the case of a "copy" instruction, the value to be written is read from another register in the register block. Bits four through seven of the instruction are fed into the first read-address input of the register block, and the desired value will appear on the first read-value output. Hence, the first read-value output of the register block is the second source of values to be written.

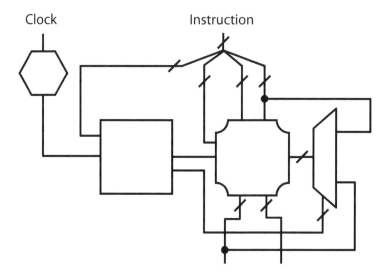

The final instruction that writes a value is the "add" instruction. In this case, the eight rightmost wires of the instruction cause two values to be read from the register block. These two values are to be added together, and the sum is the source of the value to be written.

This can be accomplished by connecting an adder to the two values read from the register block.

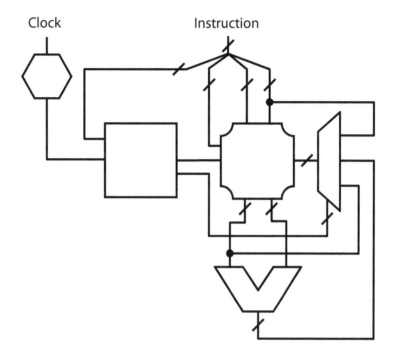

We have now accounted for every possible source of values to be stored in the register block. The wires we have just added are always connected. Regardless of which instruction is being executed, the last argument in the instruction, the first value read from the register block, and the sum of both values read from the register block will all be sent to the multiplexer. It is up to the controller to select the appropriate value to write to the register block, and to decide whether to write anything at all.

7.6 Outputting Values

Implementing the "print" operation is quite simple. The print operation is supposed to send the value from the specified register to the external output-value lines, and set the output-signal line to high voltage. The number of the register to read is specified on wires eight through eleven of the instruction. These wires are already connected to the register block's first read-address input, so the register block will read from whatever register is specified on these wires and set its value on its first read-value lines.

We can simply always send the first value read from the register block to the output-value wires. The only additional work to the done is to set an output-signal line to high voltage when a "print" operation is being carried out (so that the external device knows to pay attention to the value on the output-value lines).

As this is yet another signal that depends on the current operation, it will be the responsibility of the controller. Whenever the operation on the controller's input lines is "print", the controller will set the output signal line to high voltage.

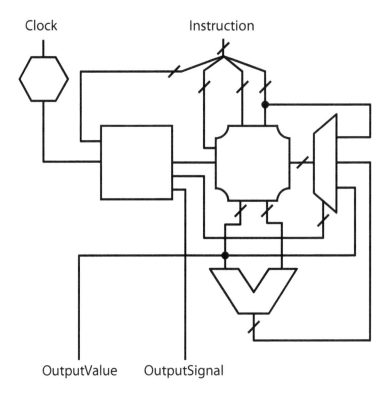

The controller should also make the output-signal line depend on the delayed clock signal being true, so the print signal line is not set to high voltage until the desired register value has finished being read. We will see this in more detail when we implement the controller.

We have now implemented everything necessary to carry out print, add, set, and copy instructions, although we must set the value on the instruction connection manually. The clock continuously cycles between high voltage and low voltage. Each time the clock changes to high voltage, we can send the next instruction on the instruction wires. Consider the following instructions to print the number ten (the sum of five and five):

```
set #3, 0x5
copy #7, #3
add #1, #3, #7
print #1
```

One possible conversion to machine code is as follows:

```
0x1305
0x2730
0x4137
0x0010
```

Passing each of these instructions to the CPU as the clock changed to high voltage would yield the following:

0x1305:

1. 0x1 (the set operation) is sent to the controller. The controller sets the register write-value multiplexer to input number zero, selecting the top input.
2. 0x3 is sent to the write-address input of the register block.
3. 0x0 is sent to the first read-address input of the register block.
4. 0x5 is sent to the second read-address input of the register block, and the top input of the write-value multiplexer. Because the top line of the multiplexer is selected by the controller, the output of the write-value multiplexer changes to 0x5 (as does the write-value input of the register block).
5. The values stored in registers zero and five appear on the read-value outputs of the register block. The value from register zero is sent to the output-value lines and the bottom input of the write-value multiplexer. Both read values are sent to the adder, and their sum is sent to the middle input of the write-value multiplexer. All of these are unrelated to the "set" operation, but do still occur.
6. A short time later, the delayed clock signal reaches the controller. Because the clock is now high voltage, and the current operation requires a write, the write-enable line on the register block is set to true by the controller. This causes the current write-value input (0x5) to be stored in the register identified by the current write-address input (0x3). Register three now contains the value five.
7. A short time later, the clock switches to low voltage, causing the write-enable line to switch to false.

0x2730:

1. 0x2 (the copy operation) is sent to the controller. The controller sets the register write-value multiplexer to input number two, selecting the bottom input.
2. 0x7 is sent to the write-address input of the register block.

3. 0x3 is sent to the first read-address input of the register block.
4. 0x0 is sent to the second read-address input of the register block, and to the top input of the write-value multiplexer.
5. The values in registers three and zero are sent to the read-value outputs of the register block. The value from register three is sent to the bottom input of the write-value multiplexer, and to the output-value lines. Because the bottom input of the write-value multiplexer is selected, the output of the write-value multiplexer changes to the value from register three (as does the write-value input of the register block). Both register read values are added, and the sum is sent to the middle input of the write-value multiplexer.
6. A short time later, the delayed clock signal arrives at the controller, causing the write-enable line of the register block to be set to true. This causes the current write-value (the contents of register three) to be written to the register identified by the current write-address (0x7).
7. A short time later, the clock switches to low voltage, and the write-enable line changes back to false.

0x4137:

1. 0x4 (the add operation) is sent to the controller. The controller selects the middle input on the write-value multiplexer.
2. 0x1 is sent to the write-address input of the register block.
3. 0x3 is sent to the first read-address input of the register block.
4. 0x7 is sent to the second read-address input of the register block, and to the top input of the write-value multiplexer.
5. The values from registers three and seven appear on the read-value outputs of the register block. The value from register three is sent to the output-value lines and the bottom input of the write-value multiplexer. Both register values are added together and the sum is sent to the middle input of the write-value multiplexer. Because the controller selected the middle input, the output of the write-value multiplexer changes to the sum of the two register values (as does the write-value input of the register block).
6. A short time later, the delayed clock signal arrives at the controller, causing the write-enable line to change to true. The sum of the two register values is stored in register one.
7. A short time later, the delayed clock changes to false, causing the write-enable line to return to false.

0x0010:

1. 0x0 (the print operation) is sent to the controller.
2. 0x0 is sent to the write-address input of the register block.
3. 0x1 is sent to the first read-address input of the register block.
4. 0x0 is sent to the second read-address input of the register block, and the top input of the write-value multiplexer.
5. The values from registers one and zero appear on the read-value outputs of the register block. The value from register one also appears on the output-value connection, as well as the bottom input of the write-value multiplexer. Both register values are sent into the adder, and their sum is sent to the middle input of the write-value multiplexer.
6. A short time later, the delayed clock signal arrives at the controller. The controller sets the output-signal connection to true, indicating that the value on the output-value lines should be observed.
7. A short time later, the delayed clock switches back to low voltage. The controller sets the output-signal line back to low voltage.

7.7 Reading Instructions

The portion of the CPU that we have created thus far can carry out almost all of the instructions, but we must input each instruction each time it is to be carried out. To truly automate the task, instructions should arrive automatically.

As is done with many types of computers, we will store our program in a memory bank that is external to the CPU. The CPU will send out the address of the desired instruction, and the memory bank (referred to as "main memory") should send back the instruction from the specified address.

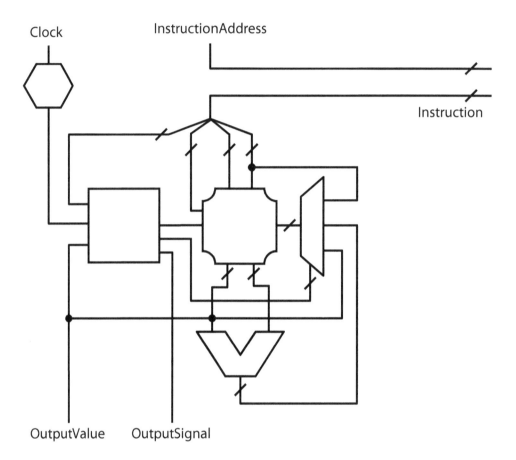

Most computers have a main memory that is very large relative to the number of the registers inside the CPU. Any device that accepts an address and outputs the value at the specified address will suffice. As we are only working with very small programs, we can simply use an external register block to store the instructions.

Zooming out for a moment, we can see how a register block containing the program could be connected to the CPU to retrieve instructions.

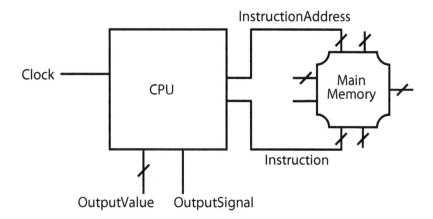

How the program is loaded into main memory will be addressed later. For now, the important property to observe is that when the CPU places an address on the outgoing instruction-address connection, the instruction at that address will appear on the incoming instruction connection. We are also assuming that the first instruction of the program begins at address zero.

We can now run the program by specifying the instruction address each time the clock switches to high voltage, rather than specifying the instruction itself. Consider the simple four-instruction program from the previous section.

```
0x1305
0x2730
0x4137
0x0010
```

To run this program, we would set the instruction-address connection to 0x0 when the clock switched to high voltage. Main memory would then output the value 0x1305 on the instruction line, and the CPU would set its register number three to a value of 0x5 (as the instruction indicates). The clock would then switch to low-voltage. As the clock switched to high voltage again, we would set the value 0x1 on the instruction address connection. This would cause the instruction 0x2730 to arrive at the CPU from main memory, and so on.

It's much simpler to specify instruction addresses instead of the actual instructions, but this is still not completely automated. We need to add some more functionality to the CPU to make it generate instruction addresses on its own. To carry out the four instructions above, we simply need the instruction addresses 0x0, 0x1, 0x2, and 0x3, as the clock switches to high voltage four times.

To accomplish this, we can add a register to hold the current instruction address on the instruction-address output of the CPU. This address should not be changed mid-way through carrying out an instruction, or else a new instruction will be read from main memory, interrupting the currently executing instruction. To keep the address of the currently-executing instruction stable, we will calculate the next address into a different register. Calculating the next address is as simple as adding one to the current address.

Adding these components yields the following:

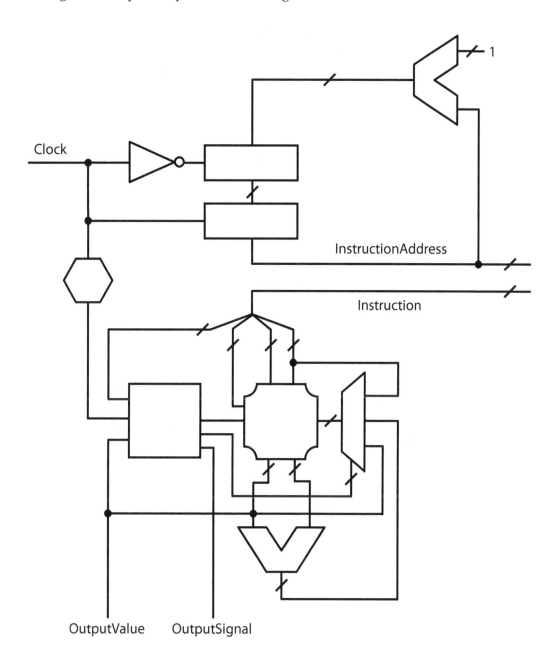

You can see the two new registers, the "next address" on the top, and the "current address" below it. The top input of the adder is always one, which can be accomplished by connecting the last wire of this input to high voltage, and all the remaining wires to low voltage. This creates the value $(0001)_2$.

When the clock signal is low-voltage, the write line on the "next address" register is set to true. This sets the "next address" register to the value of the "current address" register plus one, thanks to the adder. When the clock switches to high-voltage, the "next address" register stops accepting new values, and the "current address" register begins accepting new values. This copies the value from the "next address" register to the "current address" register. This value is then sent out over the instruction-address output of the CPU, and sent to the adder to calculate the new "next address".

The clock cycling between high and low voltage will cause the value in the current address register to increase by one each time the clock switches to high voltage. The CPU is now capable of transitioning from one instruction to the next, with no input other than the cycling clock.

7.8 Execution Flow Control

The simple adding loop at the top of the CPU can increment the current instruction from one value to the next. This is useful most of the time, but sometimes the next instruction to execute isn't at an address one higher than the current instruction's address.

The first time this situation arises is at startup. Once the program is loaded into main memory, the current instruction address must be reset to the address of the first instruction. In our case, this is zero. To facilitate this reset, one additional input – a reset connection – can be added to the CPU. When the reset line is set to high voltage, a multiplexer will force the input of the "next address" register to zero. As the clock cycles between high and low voltage, this value will be loaded into both the "next address" and "current address" registers. When the reset line is released (returned to low voltage), the instruction address will begin incrementing normally each time the clock cycles.

The reset connection and multiplexer can be added as follows:

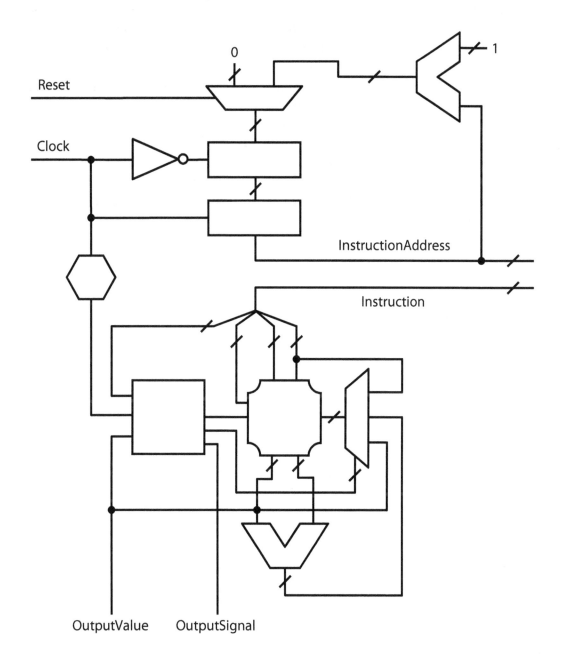

The "zero" input of the multiplexer is simply four wires connected to low voltage. This represents the binary number $(0000)_2$.

The other situation that requires a change of instruction address is the "loop" operation. As described earlier, when the operation is "loop", and the value read from a specified register is not zero, the next instruction address should be the value specified in bits zero through four of the loop instruction. The controller will need to know both when the operation is "loop" and when the first value read from the register block is zero. The controller already receives the current operation, but will also need to be connected to the first read-value output of the register block.

If the controller does see that the operation is "loop" and the value read from the register block is not zero, the next instruction address should be forced to the value in bits zero through four of the instruction. If the value read from the register block is zero, the next instruction address is calculated normally (one larger than the current instruction address). Once again, optionally forcing a value to something else is accomplished with a multiplexer.

Adding all of the connections mentioned above, as well as the new multiplexer, yields the following:

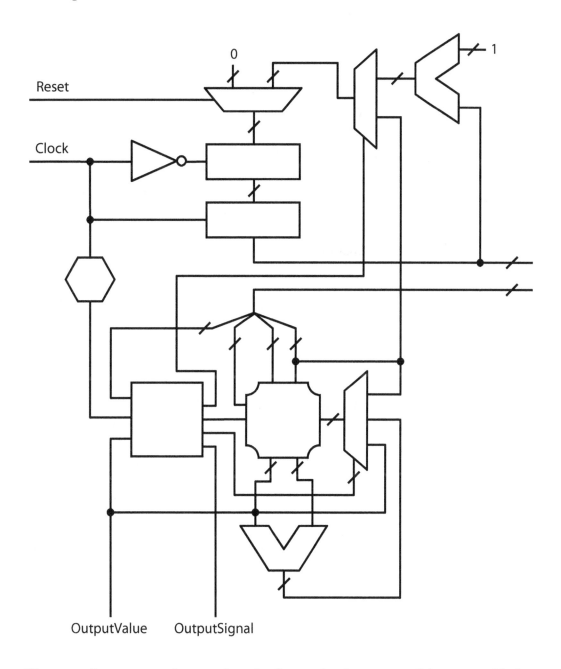

The controller can now observe when the first read-value output of the register block is zero. It can also cause the next instruction address to come from the last four bits of the current instruction, rather than the current instruction address incremented by one.

The CPU is now capable of executing all five instructions that it was intended to support. The CPU should be connected to main memory containing the program to execute, as well as a continually cycling clock. Once all of these connections have been setup, the only external action required is to hold the reset line at high voltage for at least one clock cycle, and the program in main memory will begin executing from the start.

7.9 The Controller

All of the components in the CPU have been properly interconnected. We can now implement the controller that changes how the CPU behaves for each type of operation. Looking at the CPU diagram, we can see what inputs and outputs we require. As inputs, the controller is provided with the current operation, the (delayed) clock signal, as well as the first read-value output of the register block (that is, the value stored within the register specified by bits four through seven of the current instruction).

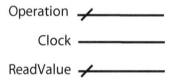

The binary representation of each operation has either exactly one bit set to one, or has all bits set to zero (in the case of the "print" operation). We can use this to our advantage by separating the four-wire operation connection into four distinct wires. We can name each wire according to the operation that sets it to a one. For example, the "add" operation sets bit number two to a one, so wire number two will be called "add". If the four-bit "add" operation is placed on the four operation wires, the "add" wire will be high voltage, and all others will be low voltage. The only exception to this is if all wires are low voltage, in which case the operation is "print".

Separating the wires gives the following:

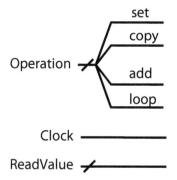

The controller must use the provided inputs to generate the correct values for each of its outputs. We will cover each output one at a time. The first output is a one-bit signal that indicates if the next instruction address should be the current instruction address plus one, or the last four bits of the current instruction. This is used to make the "loop" operation function. Setting this signal to true (or "1" or "high voltage") makes the multiplexer use the value from the rightmost four bits of the instruction as the next address. According to the definition of the operations, this should only happen when the operation is "loop" AND the value in the specified register is not zero. Another way to phrase this is: the operation is "loop" AND any one of the bits in the read-value input is a one.

We know the operation is "loop" when the "loop" input is true. Examining all of the bits in the read-value input is slightly more complicated.

We can determine if any of the read-value inputs are a one with a series of OR gates. The sixteen bits of the read value are on the left, and the output on the right indicates if at least one of the bits is one.

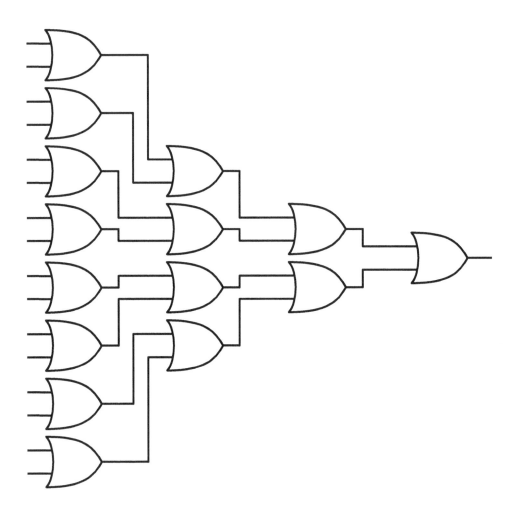

For the sake of brevity, we will draw this sixteen-input OR gate as follows:

Thus, the Operation input and ReadValue input can be used to generate the UseLoop output as follows:

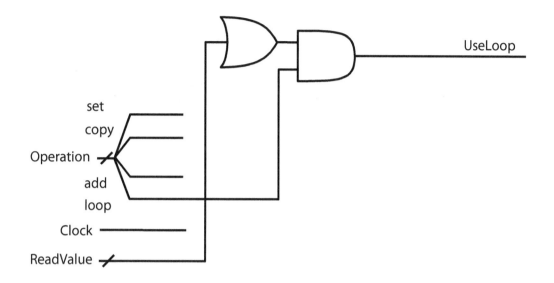

When any of the read-value bits are set to one AND the operation is "loop", the UseLoop output will be set to high voltage (causing the multiplexer to obtain the next instruction address from the rightmost four bits of the current instruction).

The next output on the controller is the register write enable. This output – RegisterWrite – is also a single-bit output. The RegisterWrite output should be high voltage when the delayed clock is high voltage, AND the operation is "set", "copy", or "add". This ensures writing is only enabled for operations that require a write, and that the write doesn't take place until the correct write address has arrived at the register block.

Adding this logic to the controller yields the following:

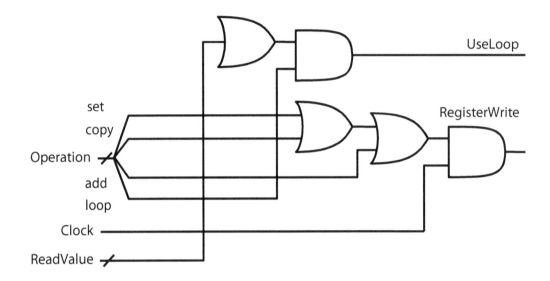

The controller will now enable register writing when the appropriate operations are ready to write to the register block.

The third output of the register block controls the origin of values to be written to the multiplexer. As the register write-value origin multiplexer has three inputs to choose from – zero, one, and two – this must be a two-bit output (to carry $(00)_2$, $(01)_2$, and $(10)_2$). We will call this output WriteOrigin.

Looking at the CPU schematic, the multiplexer should be sent a value of zero when the operation is "set" (to select the top input, and store the value from the rightmost four bits of the instruction), a value of one when the operation is "add" (to select the middle input, and store the value from the sum output of the adder), and a value of two when the operation is "copy" (to select the bottom input, and store the value read from the register block).

Operation	Selector Value
set	$(00)_2$
add	$(01)_2$
copy	$(10)_2$

To make this simpler, we can examine one bit at a time. The rightmost bit of WriteOrigin should be high voltage when the operation is "add", and zero in other cases (the other values – zero and two – have a rightmost bit of zero). We can simply connect the rightmost bit to the "add" line of the operation to achieve this, as the add line is high voltage when the operation is "add", and zero otherwise. The leftmost bit of WriteOrigin should be a one when the operation is "copy", and zero in the other cases (one and zero both have a leftmost bit of zero). The wire carrying the leftmost bit will be connected to the "copy" wire from the Operation input.

You may notice that this sets the write-value selector to zero whenever the operation is not "copy" or "add". This is the correct selector value for the "set" operation, as the multiplexer should select the top input. This behavior is also acceptable for "loop" and "print" because the write value is ignored for these operations. The write-enable connection is set to false for "loop" and "print" operations, so it does not matter what is on the write-value connection; nothing will be written.

Making this pair of connections gives the following:

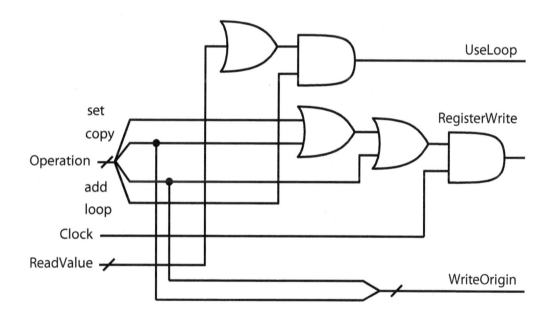

These connections create the proper values on the two-wire WriteOrigin output: zero for set operations, one for add operations, and two for copy operations. It doesn't matter what value is issued on other operations as register writing is only enabled for set, add, and copy.

The final output is the print signal output. PrintSignal should be high voltage when the operation is "print" (all four Operation wires at low voltage), and the delayed clock is high voltage. The creation of the UseLoop output showed how many OR gates can determine when at least one of multiple bits is high voltage. Following this group of OR gates with a NOT (or, equivalently, changing the last OR gate to a NOR gate) will detect when all of the bits are zero (that is, none of the bits are high voltage). The PrintSignal output is a combination of all operation bits being zero AND the delayed clock.

Implementing this with a few gates can be accomplished as follows:

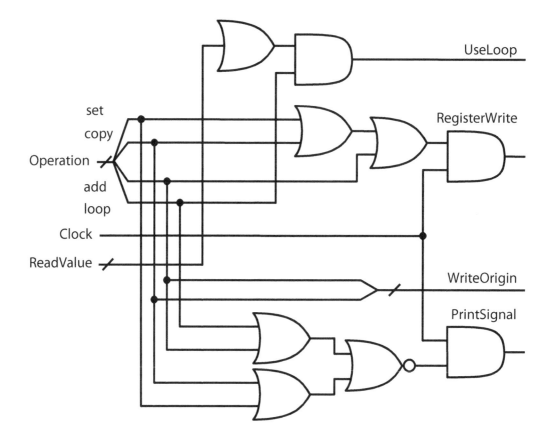

This concludes the implementation of the controller. You can now trace several instructions through the CPU schematic and this controller schematic to observe that the controller behaves correctly. Each operation should cause the controller to make the appropriate connections within the rest of the CPU.

7.10 Exercises

1. Can this CPU support a copy instruction with the same register specified as the read register and write register (e.g. 0x1330, or "copy #3, #3")? Why or why not?

2. Can this CPU support an add instruction with a write register that is the same as one (or both) or the read registers (e.g. 0x4335, or "add #3, #3, #5")? Why or why not?

Part II: Hardware Programming

8

Common Computers

8.1 x86 CPUs

The CPU in the previous section was only capable of performing relatively basic tasks. For more versatile computing, more complex processors have been developed. Consumer CPUs come in a wide variety of forms from several different manufacturers. Although the internal implementations can vary significantly, most consumer computers support the same instruction set. This instruction set is called "x86".

Each type of CPU has benefits and drawbacks. Some are faster, some use less power, and some are simply less expensive to produce. So long as each one can carry out all of the instructions in the instruction set, they can all run the same programs.

These CPUs can carry out a huge number of operations compared to the simple CPU we designed in the previous section, but still operate on the same principles. These processors have registers for temporary storage, arithmetic units for performing mathematical calculations, circuitry to determine the address of the next instruction, and connections to main memory to retrieve instructions. Building an x86 CPU is simply a matter of adding more components and connections to support the additional instructions. What makes commercial CPUs special is the use of clever design techniques that can increase execution speed.

8.2 CPU Evolution

To say that a CPU supports the x86 instruction set is not always precise. The CPUs used in typical computers were once only 16-bit devices, largely using instructions that manipulate 16-bit values. These devices can be referred to as x86-16 to be more specific. As applications grew in complexity, so did CPUs.

The x86-16 processors were followed by devices capable of manipulating 32-bit values. Along with this change came, of course, all new instructions to take advantage of the more powerful capabilities. As many users had already invested time and money into obtaining 16-bit programs that used 16-bit instructions, it was important that these new 32-bit devices continued to support the 16-bit instructions. Rather than designing an entirely new instruction set, x86-32 simply extends the x86-16 set of instructions.

To fully maintain compatibility, the new devices need to not only support the older instructions, but they also needed to behave in the same way as the old devices. When an x86-32 CPU is powered on, it operates in 16-bit mode, accepting only x86-16 instructions. If a special sequence of instructions is issued, the device switches to 32-bit mode, and can accept 32-bit instructions. This allows newer, faster 32-bit CPUs to continue running all the old 16-bit operating systems and programs. Unless these special instructions are specified, the x86-32 CPUs behave exactly as an x86-16 device would. If the user has newer 32-bit software, the new software can simply switch the CPU into 32-bit mode and take full advantage of the more powerful 32-bit instructions.

As computing continued to evolve, even the 32-bit devices became too limiting. Eventually, a natural successor emerged: x86-64. These devices add even more instructions that are capable of manipulating 64-bit numbers. Just as was seen in the switch from 16-bit to 32-bit, many users have older software that they would like to continue using on newer hardware. When an x86-64 CPU is powered on, it still begins in 16-bit mode, accepting x86-16 instructions. If the user only has software that uses the x86-16 instructions, it will run as expected. Software that uses the x86-32 instructions can switch the processor into 32-bit mode to use newer instructions. If the user has even newer software, it can switch the CPU from 32-bit mode into 64-bit mode, allowing access to all of the even newer x86-64 instructions.

By always extending the device functionality instead of replacing it, CPUs can still run programs from decades earlier. The user's software can select which level of functionality the CPU should expose. This decision is typically made by the operating system, which can then launch other programs.

8.3 System Startup

The many cells of main memory have not been explicitly set to any useful values when a computer is turned on. To carry out a program, the machine code instructions for the program must be loaded into memory so that the CPU can read them. We simply assumed that the memory contained these values when designing the sample CPU. In the emulator, the application reads these values from the instructions.txt file and places them in memory. Common computers take a different approach.

The CPU is a general-purpose device, capable of carrying out whatever programs it is issued. Alongside the CPU there is a small storage device containing a special piece of software called the BIOS, or "Basic Input/Output System". Unlike an operating system, the BIOS serves a fairly specific purpose: initializing the system and providing a basic interface to some of the hardware.

The BIOS cannot carry out applications for the user, but can communicate with the other devices within the computer, such as the CPU, main memory, and storage devices. When a computer is first turned on, the BIOS is in control. With respect to running other programs on the CPU (such as an operating system), the main job that the BIOS performs is to load a program from a specified device, and then relinquish control to this program. This initial program may be loaded from internal storage drives, from external storage media, or over a network, depending on what the BIOS supports.

This "initial" program that the BIOS loads could be a single program that the computer is meant to run, such as the Fibonacci program. The initial program could also load a larger program, such as an operating system. The operating system could, in turn, load even more programs. All that the BIOS will do is copy data into memory, from a specified device, then instruct the CPU to begin carrying out instructions from memory.

8.4 Running Instructions

The following chapters are focused on creating instructions for x86 CPUs. If you have a personal computer, you probably have a CPU capable of running x86 instructions. If this is the case, you can simply use the CPU you have. This approach does have some drawbacks, but is probably the most satisfying way to experiment with x86 machine code.

The BIOS checks several devices in search of machine code that is written in a specific format. When a device is found with properly formatted data, the BIOS copies this data into memory, then passes control to this program. The additional resources bundle provided with this book contains a utility – USBWriter – that will write properly formatted data to a USB drive. The first step to programming your x86 computer is to ensure your BIOS is capable of reading instructions into memory from this USB storage device.

We will first need to write properly formatted data to the device. Back up all data that you wish to keep from the USB drive. Any data on the device will not be readable after writing the machine code. Once the device no longer contains important data, run "write.bat" from the folder for this section – 08.04 – to write some sample machine code to the USB drive. You will have to select the drive and then press the "Write" button to write the data. Be sure you select the correct drive! The data on whatever drive you choose will no longer be accessible. You should now have a properly prepared USB device for the BIOS to read.

The BIOS is in control when you first start (or "boot") your computer. While initially starting up, there is usually a key that can be pressed to enter the BIOS setup menu (typically one of "delete", "F2", etc.). Restart your computer and enter this setup menu. Take note of any settings you intend to change, so you can restore your settings to their previous state later.

Once you have entered the setup menu, you need to locate the settings that configure the list of boot devices to check for properly formatted machine code. Follow the on-screen instructions to change this list, and add your USB drive to the list. Ensure that your USB device appears higher on the list than any other devices. If the USB device is lower on the list, the BIOS may simply find and use your normal operating system first, instead of using your code.

Note that some systems won't list the USB drive as a valid boot device unless it was plugged in before the computer was turned on. If you still cannot add your USB drive as a boot device, your BIOS may not support starting the system from a USB device.

If you have your properly formatted USB device connected to the computer, and have the BIOS configured to look for data on the USB device, you should be ready to run instructions on your x86 CPU. Restart your computer with the USB drive connected and you should be presented with a "16-bit!" message instead of your normal operating system. Remove the USB drive and restart again to return to your usual software.

You can use an x86 computer emulator if you do not have an x86 computer to use, do not have a computer capable of starting from a USB device, or simply don't want the hassle or risk of running machine code on your CPU. Running "emulate.bat" from the folder for this section will pass the machine code to an x86 emulator instead of writing it to a USB drive. You should be presented with the same "16-bit!" message.

9

x86 Architecture

9.1 Memory

The memory used in x86 systems is slightly different than the main memory used with the sample CPU from the previous section. The most noticeable difference is that x86 systems can both read from and write to main memory. The CPU still reads instructions from memory, but is also capable of carrying out instructions that read data from memory and write values back into memory.

The quantity of memory at each address is a byte (eight bits). Although there are several instructions that read and write quantities larger than a byte, the primary unit of memory is the byte. The smallest amount of memory that may be read or written is one byte, and any larger operations simply affect multiple bytes (multiples of eight bits).

Registers can manipulate values of various different sizes, but memory consists of a collection of bytes. To store values into memory from a register, the value must first be broken up into individual bytes. For example, the 32-bit number 0x12345678 would require four bytes of memory to be stored.

Reading and writing multi-byte numbers is carried out in a manner that is somewhat unintuitive. Whenever a multi-byte number is written to memory, the individual bytes are reversed. This format is referred to as "little endian", as the "little" end of the number comes first when stored in memory. Consider a 32-bit register containing the value 0x12345678. If this number was written to memory beginning at address 0x10, then the memory at address 0x10 would contain the value 0x78, 0x11 would contain 0x56, 0x12 would contain 0x34, and 0x13 would contain 0x12. The four bytes are reversed. In other words, the four bytes of memory at addresses 0x10, 0x11, 0x12, and 0x13 would contain 0x78, 0x56, 0x34, and 0x12, respectively.

This byte-reversing is simply how data is stored in memory. The CPU could later be instructed to read the 32-bit number beginning at memory location 0x10 into a register. The bytes would be loaded into the register in reverse order – 0x12 0x34 0x56 0x78 – yielding the original 32-bit number 0x12345678. Likewise, the 16-bit number 0xABCD would be written to memory as 0xCD 0xAB, and read back as 0xABCD.

As the bytes are reversed when written, then reversed again when read back, the reversal does not usually make a difference; the number is only reversed while in memory. However, the byte reversal is noticeable when the sizes of the write operation and read operation differ. If the number two hundred fifty-six is stored in a 32-bit register, the register would contain the value 0x00000100. Copying this number into memory beginning at address 0x4 would place 0x00 at address 0x4, 0x01 at address 0x5, 0x00 at address 0x6, and 0x00 at address 0x7. If the program subsequently read from memory at address 0x4 into a one-byte register, the register would receive the value 0x00, not two hundred fifty-six.

The advantage to using little-endian storage is that when a number is read into a smaller register, the lower bytes are read first. This allows larger numbers to be converted to smaller numbers automatically. For example, if the number 0xABC was placed in a 32-bit register, the register would contain 0x00000ABC. This value would be written to memory as 0xBC 0x0A 0x00 0x00. If a 16-bit value was read from the same starting address, the first two bytes would be read (and reversed). This would place the value 0x0ABC into the 16-bit register, preserving the original value.

9.2 General-Purpose Registers

A general-purpose register is any register that can be used for temporary storage and a variety of calculations. Some instructions do not allow registers to be specified, always using the same register for input or output. This automatic use does not preclude the register in question from being used for general computation at other times.

Different general-purpose registers are available in each of the x86-16, x86-32, and x86-64 modes. All general-purpose registers are of the same size in each mode: 16-bit in x86-16, 32-bit in x86-32, and 64-bit in x86-64. Some of the registers also allow smaller sections of the whole register to be used.

Consider x86-64 mode, in which each general-purpose register is 64 bits wide. One of the 64-bit registers is named "R8". In addition to using all 64 bits at once, the rightmost 32 bits can be used via the register name "R8D". The rightmost 16 bits can be accessed via the register name "R8W". Finally, the rightmost eight bits can be accessed via the register name "R8B". Although each of these has a different name, they all refer to parts of the same register. If the value 0x0123456789ABCDEF was written to the register R8, followed by the value 0x12 being written to register R8B, the register R8 would contain the value 0x0123456789ABCD12.

The conceptual layout of this particular register is as follows:

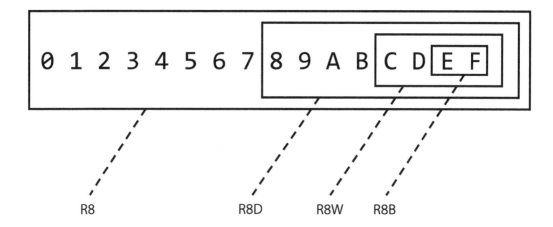

The following tables summarize the general-purpose registers that are available in each mode. Each row is a physical register, while each column is the name used to refer to a part of the register.

x86-16 (16-bit mode):

Register	Bits 0-15	Bits 8-15	Bits 0-7
0	AX	AH	AL
1	CX	CH	CL
2	DX	DH	DL
3	BX	BH	BL
4	SP		
5	BP		
6	SI		
7	DI		

x86-32 (32-bit mode):

Register	Bits 0-31	Bits 0-15	Bits 8-15	Bits 0-7
0	EAX	AX	AH	AL
1	ECX	CX	CH	CL
2	EDX	DX	DH	DL
3	EBX	BX	BH	BL
4	ESP	SP		
5	EBP	BP		
6	ESI	SI		
7	EDI	DI		

x86-64 (64-bit):

Register	Bits 0-63	Bits 0-31	Bits 0-15	Bits 8-15	Bits 0-7
0	RAX	EAX	AX	AH	AL
1	RCX	ECX	CX	CH	CL
2	RDX	EDX	DX	DH	DL
3	RBX	EBX	BX	BH	BL
4	RSP	ESP	SP		SPL
5	RBP	EBP	BP		BPL
6	RSI	ESI	SI		SIL
7	RDI	EDI	DI		DIL
8	R8	R8D	R8W		R8B
9	R9	R9D	R9W		R9B
10	R10	R10D	R10W		R10B
11	R11	R11D	R11W		R11B
12	R12	R12D	R12W		R12B
13	R13	R13D	R13W		R13B
14	R14	R14D	R14W		R14B
15	R15	R15D	R15W		R15B

64-bit mode provides the most space for computation, offering a maximum of 1024 bits of general-purpose register space inside the CPU. 32-bit and 16-bit modes provide 256 and 128 bits, respectively. 64-bit mode also provides access to the byte-sized portion of the SP, BP, SI, and DI registers, which were not available in any of the other modes.

9.3 Special-Purpose Registers

Special purpose registers are registers that are not used for general computation. The simplest special purpose register is the "flags" register. A "flag" is a bit that is used to indicate a certain property, rather than specify part of a number. For example, a byte (eight bits) could indicate whether each of eight lights are turned on or off. Each bit of the byte is a flag that indicates the "on" or "off" state; interpreting all eight bits together as a number is not very useful.

The "flags" register stores a set of flags that expose the current state of the CPU, so that it may be observed and modified by various instructions. Each bit of the value in the flags register represents a different aspect of the state. Bit number six is the "zero" flag. This bit is set (given a value of one) when the result of an operation is zero, and cleared (given a value of zero) otherwise. For example, immediately after adding the values five and seven, bit six of the flags register would be zero. After adding the values three and negative three, the flags register would have bit number six set to a one.

Some of the other commonly used special-purpose registers are the "segment registers": SS, CS, DS, ES, FS, and GS. The segment registers act as a base address for memory access. For example, if a segment register specified the base address 0x00010000, an attempt to access memory at location 0x123 using the segment register would actually access the memory at 0x00010123. The CPU reads instructions using the segment specified in the CS register. If an instruction indicated that the CPU should "jump" to the instruction at address 0x456, the CPU would actually jump to the instruction at the address indicated by CS plus 0x456.

To understand why segment registers are useful, recall the Fibonacci program from the first part of this book. The Fibonacci program always began at memory address zero, so it was safe to issue instructions such as "jump back to the instruction at address seven" at the end of a section of code to be repeated. More complex systems may want to run several programs, placing them in different memory locations. Using a segment register allows programs to run without consideration for their location in memory. Before each program is run, the segment register can be updated to point to the beginning of the program in memory. The program can then access memory as if it were at location zero in memory, and the segment register will redirect the access to a location relative to the beginning of the program.

In some CPU modes – "real" modes – the segment registers simply contain a base address. This base address automatically offsets each memory address that programs access. In other modes – "protected" modes – each segment register holds a number that identifies an entry in a table. The table contains entries that describe regions of memory. In addition to the base address where the region begins, these memory descriptor entries also contain a length.

Protected mode allows programs to be restricted to a certain region (or "segment") of memory. Whenever a program accesses memory in this mode, the base address is still added to the requested address, but the distance from the beginning of the region is also checked. If the program attempts to access memory outside the specified range, the processor raises a notification so that the issue can be dealt with. Typical responses are to either terminate the program or assign more memory to the program.

If the current selector identifies a table entry specifying a base address of 0x00010000 and a length of 0x100, memory accesses would be limited to the range 0x00010000 – 0x000100FF. The program could request a byte from memory location 0x000000A0 (which would be redirected to read a byte from memory location 0x000100A0). If the program attempted to read from memory location 0x00000700, the CPU would raise a notification indicating that the program attempted to read outside of its assigned region.

Another important register is the "instruction pointer" (or "program counter"). This register contains the address of the next instruction to be executed. Just like in the CPU from part one of this book, the CPU will automatically add to the instruction pointer to make it point to the next sequential instruction, unless an instruction makes it "jump" to a specified location. Like the general purpose registers, this register has different names at different sizes. The 16-bit, 32-bit, and 64-bit portions of the register are IP, EIP, and RIP, respectively.

The final special registers that we will use are the control registers: CR0, CR1, CR2, CR3 and CR4. These registers are used to initiate processor-oriented actions and observe which mode the processor is using. Writing different values to these registers causes different actions to occur. One common use of control registers is changing modes, between 16-bit, 32-bit, and 64-bit. This is accomplished by writing the appropriate values into the control registers.

9.4 Instruction Lengths

Unlike the CPU from part one, x86 CPUs process variable-length instructions. This means that not every instruction is the same size. This makes reading instructions more difficult as it is impossible to know how many bytes an instruction uses until part of the instruction has been read. The benefit of this design decision is that it allows for very advanced instructions, while still allowing simple instructions to use very little space.

The x86-32 instruction set includes instructions to load a specified 32-bit value into a 32-bit register. More than four bytes are required to specify the instruction type, which register to set, and the four-byte (32 bit) value to load. While this instruction is larger than four bytes, a simpler instruction such as "add the first two registers" requires only two bytes. Variable length instructions allow each instruction to occupy a minimal number of bytes.

In addition to minimizing the space required by simple instructions, allowing different instruction lengths makes it possible to have longer, very complex instructions. For example, the following line is a valid x86 instruction:

```
mov dword [gs: eax + ebx * 8 + 0x12345678], 0x0ABCDEF0
```

In English, this instruction says:

1. Use the value in register GS to look up the specified memory descriptor.
2. Start with the base address from the specified memory descriptor
3. Add the value contained in register EAX
4. Add the value contained in register EBX multiplied by eight
5. Add the value 0x12345678
6. Treat the total sum as a memory address, and write the four-byte number 0x0ABCDEF0 to memory, beginning at the calculated address

The variable length allows a great deal of information to be specified in a single instruction. This, of course, comes at the cost of space. The instruction above requires twelve bytes of memory:

```
65 C7 84 D8 78 56 34 12 F0 DE BC 0A
```

Eight of the twelve bytes are consumed simply by specifying the two four-bit numbers, 0x12345678 and 0x0ABCDEF0. The actual operation part of the instruction is only four bytes long. You can also see that instructions store values in little-endian format – the order of bytes in 0x0ABCDEF0 and 0x12345678 are reversed.

9.5 External Communication

The CPU is quite versatile, but it cannot perform many useful tasks in isolation. To facilitate integration with other pieces of hardware, the x86 architecture supports the ability to send data to external devices, as well as the ability to read data from them. Most computing systems attach a CPU to a large circuit board that contains many connections between the processor and other peripheral devices connected to the board, such as main memory.

The x86 instruction set includes instructions to send a byte of data to a specified device, as well as read a byte of data from a device. These instructions specify the "in" and "out" operations, which instruct the CPU to read and write data to and from a specified device. Although these instructions can be adequate, a more convenient method of communication has also been implemented. External hardware can interface with the main memory capabilities of the CPU, and can examine data written to (and read from) a certain range of addresses.

Communicating via this method is as simple as reading from and writing to special areas of memory. In particular, this book will make use of the special area of memory dedicated to the graphics device. The following sections display colored text by writing values to the range of memory addresses, beginning at 0xB8000, that are used for this purpose. Each colored character consists of two bytes: one to indicate which character to draw, and one to indicate which colors to use. If you wanted to write to the third character on the screen, you would write a byte to address 0xB8004 to indicate which character to display, and address 0xB8005 to specify the color.

The color byte specifies four properties.

Bit 7	Bits 6 – 4	Bit 3	Bits 2 – 0
Emphasis	Background Color	Bright Foreground	Foreground Color

The "emphasis" bits cause some systems to make the background color bright, while other systems use it to indicate that the text should blink.

The three-bit color values are interpreted as follows:

Value	Color	Bright Color
000	Black	Dark Gray
001	Blue	Light Blue
010	Green	Light Green
011	Cyan	Light Cyan
100	Red	Light Red
101	Magenta	Light Magenta
110	Brown	Yellow
111	Light Gray	White

Thus, if the value 0x29 was written to 0xB8005, the character at 0xB8004 would be light blue, with a green background. 0x29 is 00101001b, which has emphasis set to 0b (disabled), a background color of 010b (green), foreground brightness set to 1b (enabled), and a foreground color of 001b (blue).

The values written to even-numbered addresses (e.g. 0xB8002) indicate which characters to display. Each character is specified using a standard system called "ASCII", which maps a number to a character. The following grid shows a subset of the ASCII mappings with the left four bits on the top row, and right four bits on the left column. Using this table, you can see that the dollar-sign character – "$"– is specified with value 0x24. Note that 0x20 is a space, and 0x7F is not a displayable character.

	2	3	4	5	6	7
0		0	@	P	`	p
1	!	1	A	Q	a	q
2	"	2	B	R	b	r
3	#	3	C	S	c	s
4	$	4	D	T	d	t
5	%	5	E	U	e	u
6	&	6	F	V	f	v
7	'	7	G	W	g	w
8	(	8	H	X	h	x
9	)	9	I	Y	i	y
A	*	:	J	Z	j	z
B	+	;	K	[	k	{
C	,	<	L	\	l		
D	-	=	M	]	m	}
E	.	>	N	^	n	~
F	/	?	O	_	o	N/A

Another method of communication is via "interrupts". An extra set of inputs can be added to processors to trigger immediate actions. When one of these inputs changes to a certain value, the CPU looks up the appropriate handler for the interrupt, and then forces the flow of execution to jump to the handler. This effectively "interrupts" the currently running program so that the event may be dealt with. Once the interrupt has been handled, the interrupt handling code exits and the program continues on normally. Programs may also interrupt themselves by triggering an interrupt manually, causing the interrupt-handling code to run at will.

10

x86 Assembly

10.1 Assembling

x86 instructions are stored and executed as machine code, just like the instructions for the CPU from the first part of this book. Because x86 processors are significantly more complex than the sample CPU, the instruction encodings are also much more complex. Although entirely possible, we will not manually translate between x86 assembly and x86 machine code. This task has been automated by many programs called "assemblers".

The primary purpose of an assembler is, of course, to generate machine code that corresponds to each written instruction. The following is an x86 instruction, in assembly form, that sets the contents of register EAX to the value 0x12345678:

```
mov eax, 0x12345678
```

If this instruction is written in a text file and passed to the assembler, the assembler will generate the following output:

```
B8 78 56 34 12
```

You can see that the written instruction is encoded into a five-byte machine code instruction with the operation (move a value into EAX) as 0xB8, and the value 0x12345678 (in little-endian format) as an argument.

Programs often make use of pre-determined data that is stored outside of any instructions. To accommodate this, the assembler can be made to output specified data in addition to the machine-code versions of the instructions.

```
dw 0xABCD
mov eax, 0x12345678
db 0xEF
```

These lines instruct the assembler to generate the "word" (two-bytes value) 0xABCD, followed by the machine code to place the value 0x12345678 in register EAX, followed by the byte 0xEF:

```
CD AB B8 78 56 34 12 EF
```

Of course, any arbitrary series of bytes is not necessarily a valid instruction. Memory simply contains a set of bytes, and the CPU reads from the location specified by the instruction pointer - register IP/EIP/RIP – and treats the value as machine code. Directing the CPU to begin executing in the middle of arbitrary data will typically produce undesired results, as the CPU performs whatever operation the data happens to represent.

In addition to data and instructions, assembly text files can contain some information that does not end up in the output file at all. Any text that follows a semicolon is simply a comment and is ignored by the assembler. These comments are written to aid in understanding the intent of the code near them; they do not affect the generated machine code.

Instructions are encoded differently for CPUs operating in different modes. Not only are the encodings different, but some instructions are simply invalid depending on the CPU mode. For example, a 16-bit CPU (or a CPU operating in 16-bit mode) cannot process an instruction that uses the 64-bit register RAX. To properly convert instructions to the appropriate encoding and detect invalid instructions to prevent mistakes, the assembler must be told for which mode the instructions are intended. Statements in square brackets (e.g. "[BITS 32]") indicate which mode should be used. All instructions after one of these statements are encoded for the specified mode, until another mode is specified in square brackets. These statements alone do not cause any output to be written to the resulting file, they simply alter the way subsequent instructions are encoded.

Another item that appears in assembly listings but not in resulting output files is the label. In the Fibonacci program we designed earlier, a section of code was repeated by telling the CPU to loop back to instruction number seven. Although the CPU only understands numeric addresses, it's inconvenient to calculate and insert these addresses manually. Instead, a word followed by a colon indicates a label to the assembler (e.g. "Wait:"). This line alone does not result in any machine code or data, but can be referenced by machine code or data.

If an instruction wanted to jump back to the position where the "Wait:" label appeared, it could be written as "jmp Wait" to jump to the address where "Wait:" appears. This is simply a convenience for the programmer. When the assembler translates the program to machine code, the "Wait:" label does not cause any output, but its position is remembered. Whenever an instruction references the label (e.g. "jmp Wait"), the label name is replaced with the address where the label appeared (e.g. "jmp 0x7", if the Wait: label appeared after seven bytes of other data/instructions).

Consider an extension of the previous example:

```
mov eax, 0x12345678
Somewhere:
mov ebx, 0xFFFFFFFF
mov eax, Somewhere
```

Passing this to the assembler generates the following output:

```
B8 78 56 34 12 BB FF FF FF FF B8 05 00 00 00
```

Examining this machine code reveals the interpretation of the label. The first five bytes are the machine code shown earlier that places the value 0x12345678 in register EAX. Immediately following this is the machine code to place the value 0xFFFFFFFF into register EBX; the "Somewhere:" label did not result in any output. The next five bytes illustrate the translation of the label. The operation part of the instruction that indicates a value is to be moved into EAX is the same as before, 0xB8. The argument, however, is no longer the word "Somewhere", which the CPU does not understand. The argument has been replaced with the address where the "Somewhere:" label appeared – 0x00000005 – which is immediately after the first instruction (which occupies five bytes of space).

Assemblers can perform a variety of translations in addition to translating labels to addresses. Most assemblers allow you to specify values in a variety of formats. The x86 CPU always deals with binary values, but the assembler will accept values in other formats, and convert them to binary as it converts each statement to an instruction. The provided assembler can convert hexadecimal values in either notation, 0x123 or 123h. Binary values can be specified by appending a "b" to the number, such as 1010b. A value without any special markings is assumed to be decimal. The assembler will also convert characters inside single quotes to their corresponding ASCII values. For example, the value 0x57 could be loaded into register AL with the following instruction.

```
mov al, 'W'
```

You can try assembling and using programs that work on each of 16-bit, 32-bit, and 64-bit CPUs. You will need a computer with a 16-bit (or 32-bit or 64-bit) x86 processor to use the 16-bit sample. The 32-bit sample requires a 32-bit (or 64-bit) processor. Finally, the 64-bit code, of course, requires a 64-bit processor. All three versions work in the provided emulator.

The resources bundle contains a folder for this section named "10.01". Inside this folder is a folder for each of the processor modes. Running "assemble.bat" will run the assembler on the assembly file, translating it to machine code. As before, "write.bat" will write the machine code to a USB stick that you can use on a real computer, and "emulate.bat" will pass the machine code to the x86 emulator.

10.2 Moving Data

Perhaps the simplest operation a CPU performs is moving data. The x86 instruction set offers a great variety of ways to move data, with several variations of each one. Although there are some special ways to move data – reading from external devices, requesting data from the BIOS, etc. – the most common data movement is simply copying values into memory and registers.

Moving data is accomplished with the "mov" operation, followed by the destination, then the source. The name "move" may be a bit misleading. The data in question is not actually moved; the "mov" instruction simply copies data from one place to another.

Moving a value into a register is quite straightforward: specify the desired value and the register into which it should be written.

```
mov eax, 0x12345678
mov ax, 0xABCD
mov ah, 0xEF
mov al, 0x12
```

Recall that the register names RAX, EAX, AX, AL, and AH all specify different portions of the same register. With this in mind, you can see that after the previous four instructions are carried out, the 32-bit EAX register would contain 0x1234EF12.

Moving data between registers is accomplished in a similar manner, except that the source is a register, rather than a value. The only major constraint on moving data between registers is that sizes must match; "mov al, rax" doesn't make much sense, as there's no real way to fit the 64-bit value from RAX into the 8-bit AL register.

```
mov eax, 0x12345678
mov esi, eax
mov bx, si
mov dh, bl
mov cl, dh
```

At the end of the previous four instructions, the CL register would contain 0x78.

While moving data between registers is as simple as specifying which two registers should be used, moving data in memory offers a bit more flexibility. Main memory consists of individual bytes, but x86 instructions can carry out operations that read and write multiple bytes. The value "seven" can be expressed using one byte (0x07), two bytes (0x0007), four bytes (0x00000007), and so on. To accommodate the different sizes of values, it is necessary to specify which size is intended. In particular, "byte" indicates a single byte; "word" indicates two bytes (16 bits), "dword" (double word) indicates four bytes (32 bits), and "qword" (quad word) indicates eight bytes (64 bits).

Move instructions that are to access memory use a value in square brackets to indicate an address. The following instruction indicates that a byte of data – 0xAB – should be moved into memory, beginning (and ending) at address 0x12.

```
mov byte [0x12], 0xAB
```

Instead of specifying a value to write, the instruction could indicate a register whose contents should be written.

```
mov dword [0x12], eax
```

This instruction will write whatever value is contained in the register EAX into memory, beginning at address 0x12. The memory values at address 0x12, 0x13, 0x14, and 0x15 will be updated, as "dword" implies that four bytes should be written.

Reversing the order of the arguments performs the opposite operation: reading a value from memory. To read a 32-bit (4-byte) value from memory, the following instruction could be issued:

```
mov dword eax, [0x12]
```

You may recall the section on special-purpose registers noting that memory addresses are always relative to a base address. This base can be explicitly stated within the instruction.

```
mov dword eax, [fs:0x12]
```

The value in the segment register FS indicates a table entry that contains the base address to use. If the base address is 0x00000000, this instruction will read from memory beginning at address 0x00000012. If the base address is 0x00001000, this instruction will read from memory beginning at address 0x00001012. If no segment register is specified in the instruction, the DS segment register is used by default (except in the case of the general-purpose registers ESP and EBP, which use SS as a default segment register). When the CPU reads instructions to execute, it uses the segment identified by the CS register.

Instead of directly specifying an address, we can use addresses contained within registers. Writing data to the address 0x12 (relative to the base address specified by DS) could also be accomplished with the following two instructions.

```
mov ebx, 0x12
mov [ebx], eax
```

Because memory access is such a commonly used operation, the x86 instruction set provides a great deal of flexibility in specifying memory addresses. An "offset" can also be specified within the instruction. A multiple (1, 2, 4, or 8) of another register can also be used as an offset. Writing to memory beginning at address 0x12 can be accomplished in all of the following ways:

```
mov [0x12], eax
```

```
mov ebx, 0x12
mov [ebx], eax
```

```
mov ebx, 0x9
mov [ebx + 3], eax
```

```
mov ebx, 0x1
mov ecx, 0x2
mov [ebx + ecx * 4 + 3], eax
```

You can experiment with data movement in the "program.txt" file within the folder for this section (10.02). Change the instructions to suit your needs, save the file, then run "assemble.bat" to convert the assembly file to machine code. You may then use "emulate.bat" to test the program inside the emulator, or "write.bat" to copy the machine code to a USB drive you can use to boot your computer.

10.3 Mathematical Manipulation

Now that values can be moved all around, we can start actually doing something with the numbers. Both logical (true/false) and numerical (1/0) values were assigned to the high and low voltages used in part one. x86 CPUs can interpret values in both ways as well, providing both mathematical and logical instructions. Like the instructions to move data, each of the following mathematical operations can accept two registers as arguments.

The most basic mathematical operation – adding – is accomplished by using the "add" operation, and providing a destination and source register.

```
add eax, ebx
```

This instruction adds the value contained in EBX into the register EAX. That is, the values in EAX and EBX are added together, and the result is stored in EAX (overwriting whatever used to be there). Changing the operation to "sub" will subtract the value in EBX from the value in EAX (again, altering the value in EAX).

```
sub eax, ebx
```

Because adding and subtracting a value of one is such a common operation, dedicated instructions have been created to perform these tasks. Incrementing (adding one) and decrementing (subtracting one) are simple one-argument instructions.

```
inc eax
dec ecx
```

Addition and subtraction create results that are relatively close to the input values. For example, adding two numbers can never create a sum that is larger than twice the largest input number. Multiplication, however, can create results that are significantly larger than the input values. To accommodate these large results, the result of multiplication is placed into two registers, creating a result that consists of twice as many bits as each of the numbers being multiplied.

Multiplication instructions only require one argument; the output registers and one of the two input registers are fixed, and cannot be specified. The single argument specifies one of the two numbers to be multiplied. The implied input specifies the other number to be multiplied, and is always an A register (RAX, EAX, AX, or AL), depending on the size of the multiplication to carry out. The rightmost half of the result is also placed in this register. In the case of 8-bit values being multiplied, the left half of the result is placed in AH, which makes the full result in AX (as it contains both AH and AL). In all other cases, the left half of the result is placed in the D register (RDX, EDX or DX). The registers used in each case are summarized in the following table:

Size of Specified Input	Implied Input Register	Result Register(s)
8-bit	AL	AX
16-bit	AX	DX AX
32-bit	EAX	EDX EAX
64-bit	RAX	RDX RAX

A multiplication of two 32-bit values can be accomplished with the following instruction.

```
mul ecx
```

The fact that ECX is a 32-bit register implies that this is a 32-bit multiplication. The other (implied) input of a 32-bit multiplication is EAX. The result is a 64-bit value, with the left half stored in EDX and the right half stored in EAX. Using some actual values can make this a bit clearer.

```
mov eax, 0x0ABCDEF0
mov ebx, 0x12345678
mul ebx
```

0x0ABCDEF0 × 0x12345678 = 0xC379AAA42D2080. After these three instructions are carried out, EDX will contain 0x00C379AA, and EAX will contain 0xA42D2080.

In general, when the instruction "mul n" is issued, one of the following operations is carried out, based on the size of "n".

Size of n	Operation
8-bit	AX = AL × n
16-bit	DX AX = AX × n
32-bit	EDX EAX = EAX × n
64-bit	EDX RAX = RAX × n

The divide operation is almost the opposite of multiplication. A value split across two implied registers is divided by a specified value, and the result is stored in an implied register. The divide operation differs from multiplication in that it generates two result values – a quotient and a remainder.

The following table summarizes the implied registers for division.

Size of Specified Divisor	Dividend	Quotient	Remainder
8-bit	AX	AL	AH
16-bit	DX AX	AX	DX
32-bit	EDX EAX	EAX	EDX
64-bit	EDX RAX	RAX	RDX

The operations can, again, be summarized according to the size of the argument. In this case, the argument is "n" in a divide instruction "div n".

Size of n	Operation	Remainder
8-bit	AL = AX ÷ n	AH
16-bit	AX = DX AX ÷ n	DX
32-bit	EAX = EDX EAX ÷ n	EDX
64-bit	RAX = RDX RAX ÷ n	RDX

Using the numbers from the multiplication example, we can obtain the original two values that were multiplied.

```
mov edx, 0x00C379AA ; load the 64-bit value 0x00C379AAA42D2080 into
mov eax, 0xA42D2080 ;   registers EDX EAX
mov ecx, 0x12345678 ; load 0x12345678 into ECX
div ecx             ; EAX = EDX EAX ÷ ECX   and   EDX = the remainder
```

After these four instructions are carried out, EAX will contain the quotient – 0x0ABCDEF0 – which was multiplied with 0x12345678 earlier. EDX will contain the remainder, which is zero.

The resources folder for this section – 10.03 – contains some sample math in "program.txt" that you can experiment with.

10.4 Logical Manipulation

When the CPU circuitry was designed, high and low voltages were used to represent both numeric and logical values. The previous section outlined some of the basic mathematical operations, but x86 processors also provide a variety of instructions to manipulate values as logical statements (true / false). These instructions perform the same logical operations that were implemented in the logic gates (AND, OR, etc.).

A common use for the logical instructions is to manipulate individual bits. Setting specific bits to a desired value is quite cumbersome when using mathematical operations. The logical operations allow a level of precision that can't easily be obtained using instructions such as multiply and subtract.

The main difference between the x86 logical instructions and the logic gates presented earlier is that the logical instructions operate on many pairs of bits at once. Two values of equal size are provided as arguments, and the specified logical operation is carried out on each pair of bits. For example, consider the two 8-bit values shown below.

```
10100011
11001100
```

The result of performing an OR operation on these bits can be calculated by imagining the two values in each column being fed into an OR gate. Carrying out this process for each column yields the following:

```
10100011
11001100
11101111
```

This operation can easily be converted into x86 assembly.

```
mov al, 10100011b
or al, 11001100b
```

This sample also illustrates the use of binary notation in assembly. Appending a "b" to a value indicates that it should be interpreted as a base-two number.

Several logical operations can be performed via x86 instructions. Logical operations that do not have a dedicated x86 instruction can be accomplished by combining several of the provided operations (e.g. AND followed by NOT is equivalent to NAND).

```
mov al, 11001010b      ; al = 11001010b
mov bl, 0xAA           ; bl = 0xAA = 10101010b
or al, bl              ; al = 11101010b, bl is unchanged
not al                 ; al = NOT 11101010b = 00010101b
and al, 00001111b      ; al = 00010101b AND 00001111b = 00000101b
xor al, 10100011b      ; al = 00000101b XOR 10100011b = 10100110b
```

Manipulating specific bits can be accomplished quite easily. The OR operation can be used to set specific bits to a value of one.

```
or al, 00100001b
```

No matter what value is in AL, after this instruction is carried out, bit number five and bit number zero of AL will contain a one, and all other bits will remain unchanged. Similarly, the AND operation can be used to set specific bits to zero.

```
and cl, 00001111b
```

After this instruction, the four left bits of CL will all be zero, and the right four bits will be unchanged. The XOR operation is commonly used to toggle bits.

```
xor dl, 11110000b
```

When this instruction is complete, the four left bits will contain the opposite of what they contained before this instruction. The four rightmost bits will be unchanged.

In addition to combining values, the bits that make up a value can be shifted. The "shl" and "shr" operations shift the bits left and right, respectively.

```
mov al, 00101111b
shl al, 2
```

After the above instructions are carried out, register AL will contain 10111100b. If you interpret values as numbers, rather than collections of bits, you will notice that shifting effectively multiplies or divides the value; each shift left causes the value to become twice as large and each shift right causes the value to become half as large. This is a direct result of using base two.

The resources folder for this section – 10.04 – contains some sample assembly in "program.txt" that you can use to try different logical combinations.

10.5 Stacking Data

Main memory provides a large amount of storage that programs can make use of. In addition to more permanent resources, such as pictures and sounds, programs often need to store data temporarily. Temporary storage can be used for passing data to other sections of the program, or as a place to save the contents of registers while the register is used for some other task.

This type of temporary storage is so commonly used that the x86 instruction set features several instructions for making it simpler to access. Data can be conveniently stored in a form referred to as a "stack".

Values can be "pushed" onto a stack to store them temporarily. These values can then be "popped" off the stack in the reverse order; the last value to be "pushed" onto the top of the stack is the first value to be "popped" off.

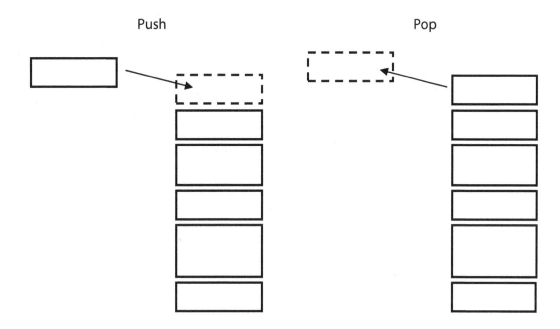

The x86 processor facilitates this action via the "push" and "pop" operations in conjunction with the "stack pointer", which is contained in the register SP/ESP/RSP. The value in the stack-pointer register is the address of the top of the stack. In other words, the stack pointer register contains the address of the last item to be pushed onto the stack. Whenever the CPU accesses the stack via the stack pointer, it uses the segment specified in the SS segment register.

Adding an item to the stack is accomplished by decreasing the stack pointer by an appropriate amount to make room for the data, followed by writing the specified data. Items may be read from the stack (without removing them from the stack) by reading from memory at the address contained in the stack-pointer register. Finally, items may be removed from the stack by increasing the value in the stack pointer register so it points to an item further down the stack.

The "push" operation adds the specified item to the stack. The "pop" operation reads a value from the top of the stack and removes the item from the stack. The following diagrams illustrate some sample instructions that use the stack. Assume that the segment register SS indicates a base address of zero.

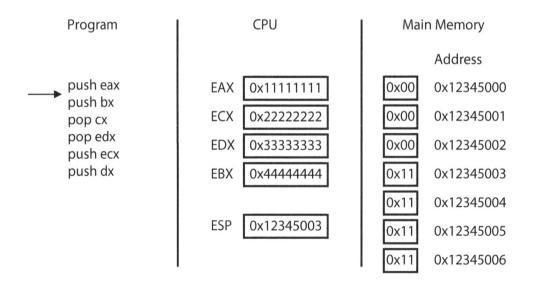

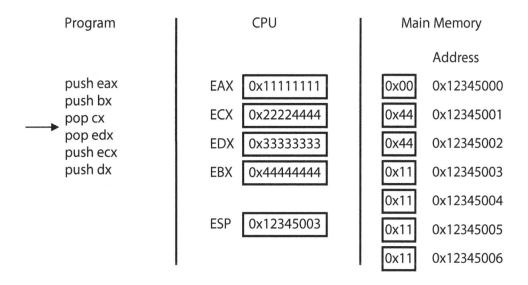

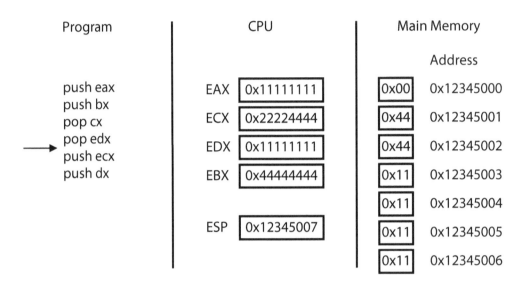

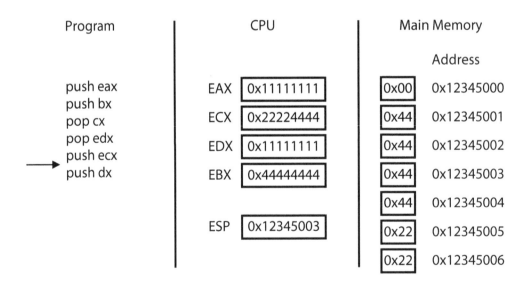

Program	CPU	Main Memory
		Address

```
          Program                    CPU                Main Memory

        push eax          EAX  0x11111111      0x00   0x12345000
        push bx
        pop cx            ECX  0x22224444      0x11   0x12345001
        pop edx
        push ecx          EDX  0x11111111      0x11   0x12345002
        push dx
                          EBX  0x44444444      0x44   0x12345003

                                               0x44   0x12345004

                          ESP  0x12345001      0x22   0x12345005

                                               0x22   0x12345006
```

As noted earlier, push and pop are provided for convenience. The same results could be obtained with other instructions.

```
push eax
```

The previous line could be replaced with the following two lines.

```
sub esp, 4
mov [esp], eax
```

The pop operation can be replaced in a similar manner.

```
pop bx
```

The previous line could be replaced with the following two lines.

```
mov bx, [esp]
add esp, 2
```

10.6 Program Control

All of the x86 instructions presented thus far simply perform an operation and lead to the following instruction. To optionally skip parts of the program, or reuse sections of code, we need to use instructions that jump to a different location in the list of instructions.

The simplest of these instructions is the "jmp" operation, which simply jumps to specified location.

```
mov eax, 0x123
jmp Somewhere
mov eax, 0xABC
Somewhere:
mov ebx, eax
```

At the end of the following code, register EBX would contain the value 0x123. The line that places the value 0xABC into register EAX is skipped by the jump instruction.

There are also a great number of conditional jumps. When certain operations are performed, the bits of the flags register are modified to store properties about the current state. One of the bits in the flags register indicates whether or not the result of the last operation was zero. The "je" and "jne" operations check this flag, and jump if the "zero" flag is set, or jump if the "zero" flag is not set, respectively.

These instructions are often used after a "cmp" operation. The "cmp" operation compares two values by subtracting one from the other, and discarding the result. This has the effect of setting the "zero" flag to true if the two values were equal (subtracting one from the other results in zero), and sets the "zero" flag to false if the values are different. A "je" – jump if equal – operation checks this flag, and jumps to the specified location if the "zero" flag is true. The "jne" operation does the opposite; "jne" jumps if the zero flag is false.

```
mov eax, 0
mov ecx, 100
AddAgain:
add eax, 5
dec ecx
cmp ecx, 0
jne AddAgain
```

The code shown above will repeatedly increase the value in EAX by five (one hundred times, to be exact). A more convenient method is to use the "loop" operation to decrease a counter (using the implied register CX), and then jump to a specified location if the decreased counter is not zero. Converting the above example to use the loop instruction produces the following:

```
mov eax, 0
mov ecx, 100
AddAgain:
add eax, 5
loop AddAgain
```

Instead of simply repeating several lines, it is often helpful to separate out a section of the program so it can be reused from several different locations. To make this task easier, the x86 instruction set provides a pair of methods to save the current position, carry out a subset of the program, then return to the original position. The "call" operation pushes the current instruction address onto the current stack, and then jumps to a specified location. This specified location can carry out the reusable set of instructions, and then issue the "ret" operation. The "ret" instruction will pop the top value off the stack, into the instruction pointer register (IP/EIP/RIP, depending on the current CPU mode), returning control to the original position that issued the "call" operation.

For example, if a program frequently needed to display the contents of registers AL, CL, DL and BL, the instructions to display these registers could be moved to a single reusable location.

```
ShowRegisters:
mov [0xB8000], al
mov [0xB8002], cl
mov [0xB8004[, dl
mov [0xB8006], bl
ret
```

Once this code is placed in the program, any other part of the program can simply use the following instruction to display the four registers:

```
call ShowRegisters
```

The "call" operation pushes the current instruction pointer address onto the stack, and then jumps to the address represented by the ShowRegisters label. Four "mov" operations carry out the requested task of displaying registers' contents. The "ret" instruction then pops the top value off the stack, into the instruction pointer register, causing the CPU to carry on executing right after the "call". The address stored on the stack is called the "return address", as it is the location to which execution will return once the "ret" instruction is issued.

This pattern of isolating reusable sections of code is quite common in many computing practices, and is referred to as a "function" or "method". Even outside of x86 assembly, the notion of "calling a function" persists, and implies that the current state of execution is saved, and a subsection of code is executed. Once the subsection of code is complete, the function "returns" to the location from which it was called. The section of code that issues the "call" instruction is referred to as the "caller", while the function being called is the "callee".

Functions often perform complex tasks, and inevitably require the use of several registers. Whatever code called the function is likely also using the registers, so the function must preserve the contents of registers before using them. Consider a function to display the contents of register AL, where the contents of the register are assumed to be between zero and nine. The function should display an ASCII character that corresponds to the value.

```
ShowNumber:
push ebx            ; The stack now contains the value that was in EBX,
                    ; followed by the return address pushed by "call".

mov bl, al          ; Copy the number to BL

add bl, '0'         ; Adding the ASCII character zero will convert the
                    ; number to a corresponding ASCII character.
                    ; e.g. the number 5
                    ; '0' = 0x30.   0x30 + 5 = 0x35   0x35 = '5'
                    ; See the ASCII table for more details.

mov [0xB8000], bl   ; Display the character

pop ebx             ; Restore the value of EBX to whatever it was
                    ; before this function was called. The return
                    ; address is now the top item on the stack

ret                 ; Return to the address on the top of the stack
```

Code may now "call ShowNumber" to display an ASCII character representing the current value in register AL. Even though ShowNumber uses the EBX register, the function saves the previous value and restores it before returning.

The previous example obtained data from a register set by the caller. To allow more data to be passed, many functions are written to accept data from the stack, rather than from registers. The caller will push data onto the stack, and then call the function. The function can then retrieve values from the stack as it needs them. Of course, when the function has completed, the values must be removed from the stack. To make this pattern simpler to implement, the "ret" instruction can be given an argument that specifies how many bytes to remove from the top of the stack after it has retrieved the return address.

ShowNumber can be re-implemented to accept a number from the stack, rather than the AL register.

```
ShowNumber:
push ebx                ; The stack now contains the value that was in EBX,
                        ; followed by the return address pushed by "call",
                        ; followed by the number to print, which was
                        ; pushed by the caller

mov bl, [esp + 8]       ; Copy the number to BL. The number is 8 bytes down
                        ; the stack, after the 4-byte EBX, and the 4-byte
                        ; return address.

add bl, '0'             ; Adding the ASCII character zero will convert the
                        ; number to a corresponding ASCII character.
                        ; e.g. the number 5
                        ; '0' = 0x30.    0x30 + 5 = 0x35    0x35 = '5'
                        ; See the ASCII table for more details.

mov [0xB8000], bl       ; Display the character

pop ebx                 ; Restore the value of EBX to whatever it was
                        ; before this function was called. The return
                        ; address is now the top item on the stack,
                        ; followed by the 1-byte number to print

ret 1                   ; Return to the address on the top of the stack,
                        ; and remove 1 extra byte after the return address
                        ; is removed.
```

This function now expects the number to have been pushed onto the stack before the "call" is issued. Printing the number five is as simple as placing it onto the stack, then calling ShowNumber. The ShowNumber function will retrieve the number from the stack to print it, and will remove the number from the stack when the function returns.

```
push byte 5
call ShowNumber
```

The resources folder for this section – 10.06 – contains some sample code that makes use of functions. Feel free to experiment with it and add your own functions.

10.7 Exercises

1. The ShowNumber function always writes the specified number to the top-left corner of the screen. Modify the ShowNumber function so that multiple calls to the function will place multiple numbers onto the screen, separated by spaces.

2. The ShowNumber function currently prints a decimal digit from 0 to 9. Modify the ShowNumber function to print a hexadecimal digit from 0 to F.

11

Program Structure

11.1 Patterns

The best way to become comfortable with programming is to practice. Many patterns are common throughout a variety of programming languages. Learning these patterns is not only useful from a design perspective, but also makes the best use of available processor functionality. Processor technology has advanced alongside programming, integrating programming patterns into hardware. Many of the operations provided by x86 CPUs are present only because they were observed to be very common practice among programmers, and thus, worth making easier.

To explore the layout of a typical program, we will develop a small application that draws triangles contained within other triangles. Specifically, it will draw triangles that alternate between black and white, and alternate between pointing up and down, with each triangle drawn inside of the last. The following samples show this process repeated four times. One large black triangle is drawn pointing up, followed by a white triangle pointing down, followed by a black triangle pointing up, etc.

11.2 Drawing Dots

The first step to drawing anything to the screen is drawing a single dot. Screens are simply a large grid of dots, and drawing an image consists of setting the color that each dot should display. Each of these dots is called a "pixel", which is a contraction of "picture element". The code for this section in the resources bundle – folder 11.02 – begins by switching the video output to an appropriate graphics mode. If you are writing this code as you read it, be sure you begin in the correct folder.

The graphics mode for this program is 640 × 480, black and white. This means the screen displays a grid of dots that is 640 pixels wide, 480 pixels high, and each dot may be either black (off) or white (on). On the screen, pixels are numbered from the top-left to the bottom-right, in (x, y) format. The pixel at the top-left corner is at location (0, 0) and the top-right pixel is at (639, 0). Similarly, location (0, 479) is the bottom-left pixel, and (639, 479) is the bottom-right.

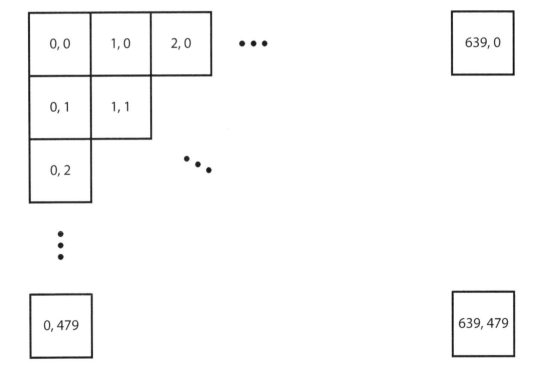

It is convenient to use two-dimensional (x, y) coordinates to describe the location of a specific pixel on the screen, but computer memory is one-dimensional; it is simply a list of bytes, not a grid. When numbering pixels in one dimension (so they can be placed in memory), the top-left pixel is number zero, the pixel to its right is number one, and to its right is number two, and so on. When the end of one row is reached, numbering continues at the left-side of the next row.

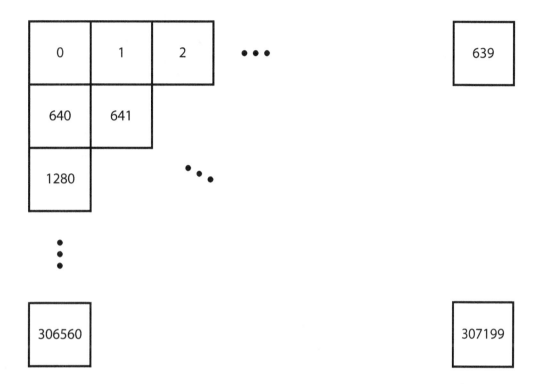

Because each pixel only has two possible colors – black or white – only one bit is needed to represent the color of each pixel. If a bit is set to one, the corresponding pixel appears white; if the bit is zero, the pixel is black.

You can imagine main memory as a list of bits, rather than bytes, to help picture the layout in memory. Pixel number zero of the image is the leftmost pixel of the top row. This is stored in the leftmost bit of the first byte, which is actually bit number seven of byte zero. Similarly, pixel nine of the image is stored in bit six of byte number one.

```
Address:         0xA0000                    0xA0001                      ...
Value:      1 0 1 0 1 0 1 0      1 1 1 1 1 1 1 1      0 0 0 ...

Byte #:            0                      1                    2  ...
Bit #:      7 6 5 4 3 2 1 0      7 6 5 4 3 2 1 0      7 6 5 ...

Pixel #:    0 1 2 3 4 5 6 7      8 9 10 11 12 13 14 15    16 17 18  ...
```

As an example, consider turning on a pixel in the top row, near the right-hand side: pixel number 638. Byte number zero contains pixels number zero through seven. Byte number one contains pixels eight through fifteen, and so on. With a bit of math, you can see that the first 79 bytes (bytes zero through 78) contain 632 bits (bits zero through 631), and hence, 632 pixels. Therefore, the leftmost bit (bit number seven) of the 80th byte (byte number 79) is pixel number 632 of the overall image. The desired pixel – number 638 – is six more bits to the right, in bit number one of byte number 79.

That calculation was pretty verbose. To carry out the calculation in a program, it will have to be converted to a sequence of mathematical and logical operations. Fortunately, the math is simpler than the description. There are eight bits in a byte, so determining how many bytes to skip (from the start of the image in memory) is as simple as dividing by eight.

$$638 \div 8 = 79, \text{ remainder } 6$$

To get to the desired pixel we need to skip over 79 bytes (groups of 8 bits), and then 6 more bits. Graphics memory begins at 0xA0000, so skipping over 79 bytes lands us at the leftmost bit of the byte at 0xA0000 + 79, or bit seven of the byte at address 0xA004F. We have now skipped as many whole bytes as needed; pixel number 638 is contained in the byte at address 0xA004F.

The remainder six indicates that the desired pixel is six more bits to the right. Setting a one six bits to the right can be obtained by starting with 10000000b, and then shifting the value to the right a distance of six, yielding 00000010b. If this byte is simply written to 0xA004F, the desired pixel will be turned on, but all other pixels contained in that byte will be turned off. We don't actually want to overwrite all eight bits of the byte, we just want to turn on the one specific bit. This is accomplished using the OR operation. Any existing bits that are turned on will stay on, and the bit in position six will be turned on, regardless of its previous value. For example:

```
01010101b OR 00000010b = 01010111b
```

Turning off a bit can be accomplished by inverting the shifted bit, then using the AND operation.

```
NOT 00000010b = 11111101b
```

```
01010111b AND 11111101b = 01010101b
```

We can now turn on and off a bit in a location specified by the pixel number (e.g. 638). This will work, but is not very convenient to use. To make drawing a bit easier, we can make the function accept more natural (x, y) coordinates, and convert to a pixel number.

Converting 2D coordinates to 1D coordinates is fairly easy. 2D numbering begins in the top left, then moves a distance down, as specified by the Y coordinate, and a distance right, as specified by the X coordinate. Comparing the 2D and 1D numbering systems shown at the start of this section, you can see that each move down a line increases the 1D coordinate by a value of 640, as there are 640 pixels on each line. Each move to the right increases the value by one. The 1D coordinate is simply:

```
640 × Y + X
```

Putting it all together, the steps to set a bit at a pixel specified in 2D (x, y) coordinates are as follows:

1. Multiply Y coordinate by 640
2. Add X coordinate to the result from step 1, this gives the pixel number (640 × Y + X)
3. Divide the pixel number by the number of bits in a byte (8). The quotient is the number of whole bytes to be skipped over; the remainder is the number of bits left to skip after all of the whole bytes have been skipped.
4. Add 0xA0000 and the quotient, this gives the address of the byte in which the pixel resides.
5. Shift 10000000b right, using the remainder as the distance to shift.
6. Combine the shifted value from step 5 with the value at the address calculated in step 4. The pixel can be turned on by using the OR operation with the shifted value, and can be turned off by using the AND operation with the inverse of the shifted value.

With a list of steps at hand, we can create the x86 instructions to generate the desired behavior. The function will accept X and Y coordinates to specify which pixel to change, as well as a one or zero to indicate whether the pixel should be turned on or off. All three arguments (X, Y, and Color) will be passed on the stack.

```
;#############################################
;
; SetPixel - Sets the value of a pixel
;
; Arguments:
;     X Coordinate (0 to 639)
;     Y Coordinate (0 to 479)
;     Color (1 = white, 0 = black)
;

SetPixel:
push eax                       ; Save current values of registers
push ecx                       ; whose values will be overwritten
push edx                       ; over the course of the function.

mov eax, [esp + 0x14]          ; eax = Y coordinate

mov ecx, 640                   ; ecx = 640

mul ecx                        ; edx eax = Y * 640

add eax, [esp + 0x18]          ; edx eax = (Y * 640) + X = pixel number

mov ecx, 8                     ; ecx = 8

div ecx                        ; eax = edx eax / ecx
                               ;     = ((Y * 640) + X) / 8
                               ; edx = remainder
mov ecx, edx                   ; ecx = edx = remainder = shift distance

add eax, 0xA0000               ; eax = eax + 0xA0000
                               ;     = ((Y * 640) + X) / 8 + 0xA0000
                               ;     = byte address

mov dl, 10000000b              ; dl = 10000000b
shr dl, cl                     ; dl = dl shr cl = shifted value

cmp dword [esp + 0x10], 1      ; determine whether we are setting
                               ; or clearing the bit
je ShowPixel                   ; If setting, go to the code to set a bit
                               ; Otherwise, continue on to clearing the bit

; Turn off a pixel
not dl                         ; Invert the shifted value
and byte [eax], dl             ; Clear the bit in graphics memory
jmp DoneSetPixel               ; Done

; Turn on a pixel
ShowPixel:
or byte [eax], dl              ; Set the bit in graphics memory

; Done changing pixel
DoneSetPixel:
pop edx                        ; Restore the original value of EDX
pop ecx                        ; Restore the original value of ECX
pop eax                        ; Restore the original value of EAX
ret 12                         ; Remove the 3 4-byte arguments
                               ; passed to the function (X, Y, value)

;#############################################
```

This function can now be called to draw some pixels on the screen. The following code will draw two adjacent pixels at (20, 10) and (21, 10), then erase the pixel at (20, 10).

```
push 20   ; X coordinate
push 10   ; Y coordinate
push 1    ; Color
call SetPixel

push 21   ; X coordinate
push 10   ; Y coordinate
push 1    ; Color
call SetPixel

push 20   ; X coordinate
push 10   ; Y coordinate
push 0    ; Color
call SetPixel
```

As you can see, arguments are pushed onto the stack in the order they are listed in the comments for SetPixel, which causes them to reside on the stack in reverse order; the last argument to be pushed on is at the top of the stack. At the very beginning of the SetPixel function, the stack contains the return address, pushed by "call", followed by the desired pixel color, followed by the Y coordinate, followed by the X coordinate. Each value is four bytes long. ESP, as always, points to the top of the stack.

Address	Value
ESP	Return Address
ESP + 0x4	Color
ESP + 0x8	Y Coordinate
ESP + 0xC	X Coordinate

The first thing the SetPixel function does is save the current values of registers that it will use. This pushes EAX, ECX, and EDX onto the stack as well.

Address	Value
ESP	Saved EDX
ESP + 0x04	Saved ECX
ESP + 0x08	Saved EAX
ESP + 0x0C	Return Address
ESP + 0x10	Color
ESP + 0x14	Y Coordinate
ESP + 0x18	X Coordinate

The remainder of the function then accesses the various stack items using the addresses shown in the table above. When the main part function has completed, it restores the original values of EAX, ECX, and EDX, using the pop instruction, which also removes the items from the stack. The "ret" instruction uses the top item from the stack, which is now the return address, to determine the location of the next instruction to be executed. In addition to removing the return address, passing the value twelve to the "ret" operation indicates that it should also remove twelve more bytes from the stack. This clears the three four-byte arguments that were passed in.

The function has now performed the desired task, returned the stack to its original state, and preserved the values in any registers it overwrote. This is what makes functions so useful; functions can be called from anywhere in a program to perform a task without destroying the current values in registers or on the stack.

The complete code for the SetPixel function is in the folder for this section; try drawing some simple images.

11.3 Drawing Lines

Some quick experimenting with the SetPixel function will reveal that drawing images one dot at a time is fairly tedious. The ability to draw horizontal lines will make drawing triangles significantly easier.

A set of horizontal lines that steadily increase in length will create the image of a triangle.

Because the lines are always horizontal, and always centered, the line-drawing function can simply accept a Y coordinate, length, and color as arguments. The function should draw a line of specified length, centered horizontally, at the height specified by the Y coordinate, in the specified color. This function will use the SetPixel function to draw each dot. SetPixel requires and X coordinate, Y coordinate, and color. The color is passed to the line-drawing function, so we can simply pass it on to each call to SetPixel. Likewise, every dot of a horizontal line is at the same height, so every dot of the line has the same Y coordinate. The Y coordinate issued to the line-drawing function can simply be passed on to each SetPixel call as well. The only value that changes along the line is the X coordinate.

The first step is to determine how a line can be centered. The screen is 640 pixels wide, so the center of the screen is at X coordinate 320 (approximately). For a line to appear centered, it should be drawn at a location such that half of the line is on the left side of the center (X coordinate 320), and half of the line is on the right. In more mathematical terms, the line should start at an X location half of its length to the left of center. More precisely, drawing left to right, the X coordinate of the line should start at:

```
320 - (length / 2)
```

Starting at this X coordinate and the constant Y coordinate, adjacent dots should be drawn until the line is of the specified width. Half of the line (length / 2) will appear on the left side of X coordinate 320, and the other half will appear on the right. This can be accomplished with the following steps.

1. Subtract the length from 640, giving:

    ```
    640 - length
    ```

2. Divide the result by two, which yields the starting X coordinate:

    ```
      (640 - length) ÷ 2
    =(640 ÷ 2) - (length ÷ 2)
    = 320 - (length ÷ 2)
    ```

3. Draw a pixel of the specified color, at the specified Y coordinate, and the current X coordinate
4. Increase the X coordinate
5. If the number of pixels drawn is less than the specified length, go back to step 3.

These steps can now be converted into x86 assembly.

```
;###########################################
;
; DrawLine - Draw a horizontal line,
;            centered horizontally
;
; Arguments
;    Length
;    Y Coordinate
;    Color (1 = on, 0 = off)
;

DrawLine:
push eax
push ecx
push edx
push ebx

mov ecx, [esp + 0x1C] ; ecx = length

mov eax, 640          ; eax = 640
sub eax, ecx          ; eax = eax - ecx
                      ;     = 640 - length

mov ebx, 2            ; ebx = 2
mov edx, 0            ; edx = 0, so edx eax = eax
div ebx               ; eax = edx eax / ebx
                      ;     = (640 - width) / 2
                      ;     = 320 - (width / 2)

mov ebx, [esp + 0x18] ; ebx = Y coordinate
mov edx, [esp + 0x14] ; edx = Color

LineLoop:

push eax              ; X coordinate
push ebx              ; Y coordinate
push edx              ; Color
call SetPixel         ; Draw the pixel

inc eax               ; Increase X coordinate
loop LineLoop, ecx    ; If there are still more pixels
                      ; to be drawn, go back to
                      ; LineLoop to draw again

pop ebx
pop edx
pop ecx
pop eax
ret 12

;###########################################
```

You can now experiment with line drawing. The following code draws a white line near the top of the screen, and then draws a black line overtop, leaving only the edges remaining. Two solid white lines are also drawn.

```
push 500        ; Length
push 30         ; Y Coordinate
push 1          ; Color
call DrawLine

push 400        ; Length
push 30         ; Y Coordinate
push 0          ; Color
call DrawLine

push 500        ; Length
push 20         ; Y Coordinate
push 1          ; Color
call DrawLine

push 400        ; Length
push 40         ; Y Coordinate
push 1          ; Color
call DrawLine
```

11.4 Drawing Triangles

With the help of the pixel-drawing and line-drawing functions, drawing a triangle is significantly simpler. Just as a line was created by repeatedly drawing dots, a triangle can be created by repeatedly drawing lines. The image of a triangle pointing up can be created by drawing lines under each other, each longer than the last. A triangle pointing down must draw longer lines above the last.

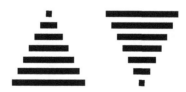

In general, a triangle can be created by starting with a line of length of one, then repeatedly drawing longer lines. Each subsequent longer line must be at a Y coordinate that is one greater than the last, for a triangle pointing down. For a triangle pointing up, the Y coordinate must be one less than the previous line. Rather than have two different functions to draw the triangles, we can simply store the "direction" as either a one or a negative one. This "direction" value can then be added to the current Y coordinate to get the next Y coordinate.

The triangle-drawing function will accept a Y coordinate for the point, a Y coordinate for the base, and a color. The function should start with a line of length one at the point Y coordinate, then draw increasingly long lines until it has reached the base Y coordinate. The detailed steps are as follows.

1. Determine if the triangle is pointing up or down. If pointing down, go to step 2. If pointing up, go to step 3.
2. Moving from point to base has decreasing Y values, so the "direction" is -1. The point is below the base, which means the point has a larger Y coordinate. Calculate the height of the triangle as follows, then go to step 4:

    point Y coordinate - base Y coordinate + 1

3. Moving from point to base has increasing Y values, so the "direction" is 1. The point is above the base, which means the base has a larger Y coordinate. Calculate the height of the triangle as follows, then go to step 4:

    base Y coordinate - point Y coordinate + 1

4. Set the current length to 1.
5. Draw a line of the current length, at the current Y coordinate, in the specified color.
6. Add "direction" to the current Y coordinate to move towards the base
7. Add 1 to the current length, to make the next line longer
8. If the number of lines drawn is less than the height (calculated in step 2 or 3), go back to step 5.

The final step, as usual, is converting to assembly.

```
;############################################
;
; DrawTriangle - Draw a triangle of pixels
;
; Arguments
;     Y Coordinate of point (0 to 639)
;     Y Coordinate of base (0 to 639)
;     Color (1 = on, 0 = off)
;

DrawTriangle:
push eax                ; Save registers
push ecx
push edx
push ebx
push esi
push edi
```

```
mov eax, [esp + 0x24] ; eax = point Y coordinate
mov ebx, [esp + 0x20] ; ebx = base Y coordinate
mov esi, [esp + 0x1C] ; esi = color

cmp eax, ebx          ; See if the point is above
                      ; or below the base
jl PointUp            ; point Y < base Y means
                      ; the point is above
                      ; the base.

                      ; Otherwise, point is below
                      ; the base

; Point is down
mov ecx, eax          ; ecx = eax = point Y
sub ecx, ebx          ; ecx = ecx - ebx
                      ;     = point Y - base Y
                      ;     = height - 1
mov edx, -1           ; edx = direction = -1
                      ; have to decrease Y
                      ; to move from point to base
jmp Draw              ; Start drawing

; Point is up
PointUp:
mov ecx, ebx          ; ecx = ebx = base Y
sub ecx, eax          ; ecx = ecx - eax
                      ;     = base Y - point Y
                      ;     = height - 1
mov edx, 1            ; edx = direction = 1
                      ; have to increase Y to
                      ; move from point to base
                      ; Start drawing

; Draw
Draw:
inc ecx               ; ecx = ecx + 1
                      ;     = height - 1 + 1
                      ;     = height
mov ebx, 1            ; ebx = 1 = line length

DrawLoop:
push ebx              ; Length
push eax              ; Y coordinate
push esi              ; Color
call DrawLine         ; Draw the line

add eax, edx          ; Move towards the base
add ebx, 1            ; Increase line length
loop DrawLoop, ecx    ; Keep drawing if not done

DoneDrawTriangle:
pop edi               ; Restore register values
pop esi
pop ebx
pop edx
pop ecx
pop eax
ret 12                ; Return, and remove arguments
                      ; from the stack
;#########################################
```

The program is now capable of drawing triangles. The code below will draw a solid white triangle, pointing up, from the top of the screen to the bottom.

```
push 0             ; Point (at the top of the screen)
push 479           ; Base (at the bottom of the screen)
push 1             ; Color (white)
call DrawTriangle  ; Draw a triangle
```

11.5 More Triangles

The program thus far is capable of drawing triangles of any height, facing up or down, of a specified color. Unfortunately, the function just draws one triangle; the image we wanted to create had many triangles. To quickly convert the program to draw many triangles, we can use a technique called "recursion", which is a situation in which a function calls itself.

In our case, DrawTriangle can call itself. The main program will call DrawTriangle to draw a large triangle, which will call DrawTriangle to draw a smaller triangle, which will call DrawTriangle to draw an even smaller triangle, and so on. Of course, if DrawTriangle always calls itself, the function will never return. In fact, because each call to the function would call itself again before returning (which clears its arguments and return address from the stack), more and more data will get pushed onto the stack until there is no more room for the stack to grow. This problem is called a "stack overflow", and is commonly caused in cases of uncontrolled recursion. As the function continues to call itself in an infinite loop, more and more arguments, return addresses, and saved registers get pushed onto the stack each time. This will continue until there is simply insufficient memory to continue.

To prevent the program getting stuck in infinite recursion and eventually running out of memory, the first change to the DrawTriangle function will be to avoid calling itself when the recursion should stop. In our case, we will be drawing smaller and smaller triangles. There is no point in trying to draw triangles with a height of zero (or smaller), so this seems like a good point to exit.

The next issue to address is what arguments should be passed to the recursive call to DrawTriangle. The answer lies in the image we wish to create. DrawTriangle requires a color, a point Y coordinate, and a base Y coordinate.

The picture below shows that each triangle's point is at the previous triangle's base. You can also see that each triangle's base is at the middle of the previous triangle's height. The final argument, color, is simply the opposite of the previous triangle's color.

The following code causes DrawTriangle to call itself recursively until the height reaches zero. Each triangle draws itself, and then draws a smaller triangle with a point at the current triangle's base, a base at the middle of the current triangle, and the opposite color compared to the current triangle. The lines in bold are new.

```
;###########################################
;
; DrawTriangle - Draw a triangle of pixels
;
; Arguments
;     Y Coordinate of point (0 to 639)
;     Y Coordinate of base (0 to 639)
;     Color (1 = white, 0 = black)
;

DrawTriangle:
push eax                 ; Save registers
push ecx
push edx
push ebx
push esi
push edi

mov eax, [esp + 0x24] ; eax = point Y coordinate
mov ebx, [esp + 0x20] ; ebx = base Y coordinate
mov esi, [esp + 0x1C] ; esi = color
```

```
        cmp eax, ebx              ; See if the point is above
                                  ; or below the base
        je DoneDrawTriangle       ; equal, height is zero, exit now
        jl PointUp                ; point Y < base Y means
                                  ; the point is above
                                  ; the base.
                                  ; Otherwise, point is below
                                  ; the base

        ; Point is down
        mov ecx, eax              ; ecx = eax = point Y
        sub ecx, ebx              ; ecx = ecx - ebx
                                  ;     = point Y - base Y
                                  ;     = height - 1
        mov edx, -1               ; edx = direction = -1
                                  ; have to decrease Y
                                  ; to move from point to base
        mov edi, ebx              ; save the top Y coordinate (the base)
        jmp Draw                  ; Start drawing


        ; Point is up
        PointUp:
        mov ecx, ebx              ; ecx = ebx = base Y
        sub ecx, eax              ; ecx = ecx - eax
                                  ;     = base Y - point Y
                                  ;     = height - 1
        mov edx, 1                ; edx = direction = 1
                                  ; have to increase Y to
                                  ; move from point to base
        mov edi, eax              ; save the top Y coordinate (the point)
                                  ; Start drawing
        ; Draw
        Draw:
        inc ecx                   ; ecx = ecx + 1
                                  ;     = height - 1 + 1
                                  ;     = height
        mov ebx, ecx              ; ebx = ecx = height
        shr ebx, 1                ; ebx = ebx / 2 = height / 2
        add edi, ebx              ; edi = edi + ebx
                                  ;     = top Y + height / 2

        mov ebx, 1                ; ebx = 1 = line length

        DrawLoop:
        push ebx                  ; Length
        push eax                  ; Y coordinate
        push esi                  ; Color
        call DrawLine             ; Draw the line

        add eax, edx              ; Move towards the base
        add ebx, 1                ; Increase line length
        loop DrawLoop, ecx        ; Keep drawing if not done

        mov ebx, [esp + 0x20]     ; ebx = base Y coordinate
        xor esi, 1                ; switch to the other color
                                  ; 1 XOR 1 = 0
                                  ; 0 XOR 1 = 1

        push ebx                  ; point = current base
        push edi                  ; base = middle of current triangle
        push esi                  ; color = opposite of current color
        call DrawTriangle         ; draw the smaller triangle
```

```
DoneDrawTriangle:
pop edi                    ; Restore register values
pop esi
pop ebx
pop edx
pop ecx
pop eax
ret 12                     ; Return, and remove arguments
                           ; from the stack

;############################################
```

The code now creates the desired image. The following call will create the image as large as possible, taking up the whole screen.

```
push 0              ; Point (at the top of the screen)
push 479            ; Base (at the bottom of the screen)
push 1              ; Color (white)
call DrawTriangle   ; Draw a triangle
```

11.6 Exercises

1. The DrawLine function performs a division by two using the "div" instruction. What is a simpler way to divide a number by two, without using the "div" instruction? Remember, all numbers inside the CPU are stored in binary format.

2. The SetPixel function divides a number by eight using the "div" instruction. The "div" instruction also generates the remainder after the division, which SetPixel makes use of. How can this remainder be found if the "div" instruction is no longer being used?

12

Operating Systems

12.1 Startup

We have created code that runs on an x86 CPU. If you used a USB drive to boot your computer with the provided code, you will have observed that the machine was indeed carrying out the specified instructions. The program has full access to main memory, and can do whatever it pleases with the CPU and other hardware. This simple arrangement may be convenient for single-purpose computers, but modern CPUs are capable of much more; complex CPUs can be used to run a great variety of programs.

Unfortunately, the move from one program to multiple programs makes things a bit more complicated. No longer can each program monopolize the CPU, or write to whatever memory it pleases; each program must be restricted in such a way as to not interfere with any other programs in the system.

The following sections will outline the design of a very simple 32-bit system that is capable of sharing the CPU between two programs, as well as restricting their access to memory. Before any of the major tasks are performed, some basic setup must be completed. When the computer is turned on, the BIOS loads a block of data from the boot device and places it in memory starting at address 0x7C00. The CPU is then instructed to begin execution at 0x7C00, and the loaded program assumes control.

```
[ORG 0x7C00]

[BITS 16]

; Disable interrupts
cli

; Enter color mode
mov ax, 3
int 0x10

; Wait forever
Wait:
jmp Wait

; Pad to 512 bytes
TIMES 510-($-$$) DB 0
DW 0xAA55
```

The first two lines are directions for the assembler, and are not translated into instructions. The "ORG" line indicates that all subsequent code should be generated relative to the address 0x7C00. Some instructions include addresses, and these addresses must take into account the fact that the BIOS placed the code at 0x7C00. For example, imagine a label "Somewhere:" appearing five bytes into the program. If a later instruction was "jmp Somewhere", the assembler should translate it to the machine code to jump to 0x7C05, not 0x0005. The next directive – [BITS 16] – causes the assembler to generate instructions for 16-bit mode, as the CPU always starts in 16-bit mode first.

The next few lines are actual instructions. The CPU receives interrupt signals from various sources which interrupt program execution and jump to a specified handler. We have not yet specified any handlers, so interrupts should remain disabled. The "cli" instruction indicates that interrupts should be ignored.

The next two lines switch the video card into colored-text mode. We have not created any drivers to communicate with specific video cards, but the BIOS is capable of basic hardware communication. Triggering interrupt 0x10 causes the BIOS to change to a video mode specified in the register AX. A value of three indicates 80 × 25 colored text mode. This mode requires two bytes per character; one byte specifies the character, the next specifies the coloring. The screen holds 25 lines with 80 characters each, beginning at address 0xB8000 in memory. Interrupts will be covered in more detail later in the chapter.

The final instruction simply causes the program to pause execution indefinitely. The last two lines of the listing direct the assembler to write the value zero until the file is 510 bytes in size, then write the two-byte value 0xAA55. Before the BIOS loads the 512-byte block of data from the chosen boot device, it first checks that the data is marked as valid executable data. This marking is simply the two-byte value 0xAA55 in the 511th and 512th bytes.

12.2 Enabling Large Addresses

To maintain compatibility with CPUs dating far back in computer history, x86 processors start up in the original 16-bit mode. The original processor that this mode imitates – the 8086 – has twenty address lines, numbered zero through nineteen. If an address larger than twenty bits is specified, the request simply "wraps" around, as any bits above twenty are discarded. Crafty programmers noticed this behavior and took advantage of it by using addresses slightly larger than twenty bits (one megabyte) to access the beginning of memory. For example, when a program accesses address 0x100123, only the rightmost twenty bits are sent out on the memory-address wires; any other bits have no effect. This causes a memory access at address 0x00123, not 0x100123. So long as programmers knew about this behavior, making use of it was convenient and reliable.

Eventually, of course, there emerged a successor to the 8086 line of chips that could handle more than one megabyte of memory: the 80286. The 80286, or "286", featured twenty-four address lines to facilitate access to sixteen megabytes of memory. This caused problems for programs that assumed addresses larger than one megabyte would wrap back around to address zero; 21-bit addresses now pointed to additional memory, not the beginning of memory. To retain compatibility with all of the old software that depended on bit number twenty being ignored, the address line for bit twenty, or simply "A20", is initially disabled on systems that use an 80286 processor or one of its successors.

You can probably imagine how the address line could be enabled and disabled. Address line twenty on the main memory unit could obtain its input from an AND gate. The inputs of this AND gate would come from address line twenty on the CPU, and a one-bit register that stores a value indicating whether line twenty is enabled or not. This scenario would cause address line twenty of the main memory unit to carry a value of one only if the CPU placed a one on line twenty, and if line twenty is enabled.

Unfortunately, both the memory and CPU were already created and provided to the computer manufacturers. Manufacturers needed to use a device outside of the CPU and memory to perform this logic. At the time, the best candidate for this task was the keyboard controller chip. Although it may seem strange, commands must be sent to the keyboard controller chip for address bit number twenty to function. Without sending the appropriate commands, all memory accesses will be "missing" bit number twenty of the address.

The startup program can be extended to also enable the A20 line.

```
[ORG 0x7C00]

[BITS 16]

; Disable interrupts
cli

; Enter color mode
mov ax, 3
int 0x10

; Enable A20
call WaitForKeyboardController
mov al,0xD1
out 0x64,al
call WaitForKeyboardController
mov al,0xDF
out 0x60,al

Wait:
jmp Wait

WaitForKeyboardController:
mov cx, 1000                        ; Set retry count
CheckKeyboardController:
in al, 0x64                         ; Read from the keybaord controller
and al, 0x2                         ; See if bit 1 is set, indicating
                                    ; that the controller is ready.
loopnz CheckKeyboardController      ; If not, and the counter is above
                                    ; zero, try again
ret

; Pad to 512 bytes
TIMES 510-($-$$) DB 0
DW 0xAA55
```

The A20 line is enabled by sending two specific bytes to the keyboard controller that instruct it to make the connection. Data can be sent to external devices using the "out" operation, and data can be read from external devices using the "in" operation. Each one accepts an address that identifies the device to access, as well as either a value to write, or a register to read into.

The value retrieved from device address 0x64 will have bit number one set to a value of one when the keyboard controller is ready for data. The WaitForKeyboardController function reads from this location (up to 1000 times) until bit number one is set. Once the controller is ready for data, the two-byte sequence to enable the address line is sent. The meaning of each bit of the values written to the keyboard controller is detailed in the keyboard controller chip specifications, and is beyond the scope of this book.

12.3 Memory Segments

All of the memory address lines are now connected. The next step of system startup is to setup how that memory will be accessed. The segment registers – CS, DS, ES, FS, GS, and SS – select an entry from a table that indicates how memory accesses should behave. Each table entry specifies a base address that offsets any addresses specified in an instruction, as well as a memory range that limits access to a certain region. Before we can begin using the table entries in the segment registers, we must create the table and instruct the processor to make use of it. This table is referred to as the "global descriptor table", or simply "GDT".

Loading a GDT requires two data structures. The first structure simply indicates where the table resides in memory, and how much space it occupies. This structure – the "GDT pointer" – contains a 16-bit value that is one less than the size of the table, as well as a 32-bit value that contains the memory address of the table.

The actual GDT entries are a bit more complex. Each table entry is an eight-byte value, and contains a 32-bit segment base address, a 20-bit segment limit (one less than the desired size), eight bits of access flags, and four bits of size flags. The values are arranged in a somewhat unintuitive manner, shown below. Each block contains four bits.

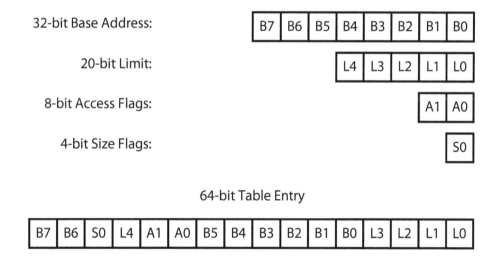

32-bit Base Address:	B7	B6	B5	B4	B3	B2	B1	B0
20-bit Limit:				L4	L3	L2	L1	L0
8-bit Access Flags:							A1	A0
4-bit Size Flags:								S0

64-bit Table Entry

| B7 | B6 | S0 | L4 | A1 | A0 | B5 | B4 | B3 | B2 | B1 | B0 | L3 | L2 | L1 | L0 |

Each bit of the access flags and size flags control a specific aspect of the memory segment. We will only use two different values for each of the access and flags fields. Bit three of the access flags specifies whether a segment contains code or data. A value of one indicates a code segment (executable, but not writeable), while a value of zero represents a

185

data segment. With other bits set to the desired settings, toggling bit number three yields 0x9A for code segments and 0x92 for data segments.

Similarly, the only setting that we will change in the size flags is the granularity flag, which is bit number three. The granularity flag indicates whether the limit should be treated as a number of bytes or a number of four-kilobyte blocks. Keeping the other desired values the same, four-kilobyte granularity is used when the size flags has a value of 0xC, and one-byte granularity is used when the size flags has a value of 0x4.

We can now setup several memory segments using the flag values and table entry layout shown above.

```
[ORG 0x7C00]

[BITS 16]

; Disable interrupts
cli

; Enter color mode
mov ax, 3
int 0x10

; Enable A20
call WaitForKeyboardController
mov al,0xD1
out 0x64,al
call WaitForKeyboardController
mov al,0xDF
out 0x60,al

; Load GDT
mov ax, 0x0000
mov ds, ax
lgdt [GDTPointer]

; Enable protected mode
mov eax, cr0
or   al, 1
mov cr0, eax

; Enter 32-bit protected mode
jmp 0x0010:ProtectedMode

WaitForKeyboardController:
mov cx, 1000                    ; Set retry count
CheckKeyboardController:
in al, 0x64                     ; Read from the keybaord controller
and al, 0x2                     ; See if bit 1 is set
loopnz CheckKeyboardController  ; If not, and the counter is above
                                ; zero, try again

ret

[BITS 32]
ProtectedMode:
```

```
; Wait forever
Wait:
jmp Wait

GDTPointer:
dw 5 * 8 - 1
dd GDT

GDT:
dq 0x0000000000000000 ; 0x0000 - Null
dq 0x00CF92000000FFFF ; 0x0008 - OS Data
dq 0x00CF9A000000FFFF ; 0x0010 - Code
dq 0x0040920B80A0077F ; 0x0018 - Program A Data
dq 0x0040920B8820077F ; 0x0020 - Program B Data

; Pad to 512 bytes
TIMES 510-($-$$) DB 0
DW 0xAA55
```

This code creates four memory segments. The first segment is not used, as a selector value of zero indicates an error. The remaining segments are summarized in the table below. Recall that the limit is one less than the size, and specifies a multiple of the granularity. In other words, the size can be calculated as:

$$size = (limit + 1) * granularity$$

Entry	Type	Base Address	Limit	Granularity	Size
0x00CF92000000FFFF	Data	0x00000000	0xFFFFF	4 KB	0x100000000
0x00CF9A000000FFFF	Code	0x00000000	0xFFFFF	4 KB	0x100000000
0x0040920B80A0077F	Data	0x000B80A0	0x0077F	1 byte	0x000000780
0x0040920B8820077F	Data	0x000B8820	0x077F	1 byte	0x000000780

Ignoring the invalid segment, the first two segments provide code and data access to the entire range of memory; they begin at 0x00000000 and have a size of 0x100000000, providing access to the full range of 0x00000000 – 0xFFFFFFFF. These segments do not offset accesses, or limit access to a certain range; a memory request for address 0x00001230 will access memory at location 0x00001230. The following two segments are significantly more restricted.

The memory segments for Program A and Program B have been setup to give each program access to approximately half of the memory that maps to the display. The memory segment for Program A begins at 0xB80A0. Because colored-text video memory begins at 0xB8000, this location is 0xA0 bytes into video memory. Each colored character

187

requires two bytes, so this is 0x50 (80) characters past the beginning of video memory. The screen is 80 characters wide, so the first line contains characters 0 – 79. This places memory location 0xB80A0 at the very beginning of the second line. When this segment is selected, accessing address 0x00000002 will be redirected to access 0x000B80A2. The size of 0x780 shows that this memory segment is 1920 bytes long. 1920 bytes holds 12 lines of 80 2-byte characters. This segment allows access to memory addresses 0x00000000 – 0x0000077F, and redirects them to physical memory locations 0x000B80A0 – 0x000B881F. Any memory requests outside this range will result in an error.

Program A's memory segment begins after the first line of text, and encompasses 12 of the 25 lines on the screen. Similarly, the segment for Program B begins after the first 13 lines of text, and also contains 12 of the 25 lines of text. Program A can access video memory that maps to lines 1 – 12, and Program B can access lines 13-24. Neither program can access line zero. Any attempt to access memory outside these ranges will generate an error.

With the table created in memory, all that remains to be done is to issue a load instruction. The load instruction is passed the GDT pointer as an argument, indicating the location of the table, and (one less than) the size. With the table loaded, we can now switch to 32-bit mode. The appropriate bit is set in the control register – CR0 – to enable 32-bit mode, and then a special "jmp" instruction is issued. This jump indicates an address as well as a segment to use for reading the program from memory. We simply use segment 0x10, which is the code segment that doesn't alter addresses at all (0x00000123 maps to physical address 0x00000123). After this jump, the CPU is in 32-bit mode and the CS register will contain 0x10. The value 0x10 in the CS register instructs the CPU to read subsequent instructions using the memory segment described by the entry 0x10 bytes into the global descriptor table. The "[32 BITS]" directive causes the assembler to begin generating 32-bit code.

12.4 Task Switching

One of the major advantages of using an operating system instead of a simple program is the ability to run multiple tasks at once. Each program is allowed to run for a short amount of time, and then paused while another program gets a chance. Each running program will make use of general-purpose registers, some memory for storing data, a stack for instructions like push, pop, call, and ret. The CPU does not have duplicate registers for each process, so running multiple programs means saving the values one program is using before switching to another. Each program will have its own stack, so we can simply push all of the CPU's state onto the program's stack before switching to another program.

All paused programs will have saved their current state saved on their stack. Once the currently running program pushes all of its CPU state onto the stack, all programs will have their state on the top of their stack. At this point, switching tasks is as simple as putting the stack pointer for the chosen program into the ESP register, and then popping all of the saved state off of the stack into the appropriate CPU registers. Whichever task has its stack pointer loaded into ESP will have all of its state restored when the data is popped back off the stack. The only thing required to select a task is to select its stack pointer prior to everything being popped off the stack.

The OS will switch between two programs. Each program will write 960 (0x3C0) letters to the screen, and repeatedly change their color. Program A will write the letter A, while program B will write the letter B. Each program will write the characters starting at address zero. Because each program has a different memory segment, the actual memory accesses will be directed to different places. The memory segment assigned to Program A resides in the top half of video memory, while the memory for Program B resides in the bottom half.

```
ProgramA:
mov eax, 0          ; Start writing at address 0x00000000
mov ecx, 0x3C0      ; Write 0x3C0 characters
LoopA:
mov byte [eax], 'A' ; Set the character to 'A'
inc eax             ; Move to the color byte
inc byte [eax]      ; Change the color
and byte [eax], 0x7F ; Remove the blink bit
inc eax             ; Move to the next character
loop LoopA, ecx     ; Keep going if not done
call SwitchTasks    ; Switch to another task
jmp ProgramA        ; Keep running forever

ProgramB:
mov eax, 0          ; Start writing at address 0x00000000
mov ecx, 0x3C0      ; Write 0x3C0 characters
LoopB:
mov byte [eax], 'B' ; Set the character to 'B'
inc eax             ; Move to the color byte
inc byte [eax]      ; Change the color
and byte [eax], 0x7F ; Remove the blink bit
inc eax             ; Move to the next character
loop LoopB, ecx     ; Keep going if not done
call SwitchTasks    ; Switch to another task
jmp ProgramB        ; Keep running forever
```

You can see that both programs make use of the same registers, illustrating the importance of saving all of the CPU registers to the stack before switching tasks.

Programs need to be able to be paused while other programs are given a chance to run. Before resuming a program, the values in the CPU registers need to be restored to the values they contained when the program was paused. If values in registers suddenly changed part way through a program because another program overwrote them, the program would likely fail quite quickly.

As you can see, both Program A and Program B work for a while, then call SwitchTasks. After a set number of calls to SwitchTasks, the current program is paused and the other progam is given a chance to run. SwitchTasks uses a counter, stored in memory location 0x8000, to count the number of times it has been called. After 90 calls, it performs a task switch and resets the counter to zero.

The "call" instruction pushes the current point of execution – the address in EIP – onto the stack. In addition to this register, we also need to save the flags register, all of the general purpose registers, and the DS register that redirects the program's memory accesses. Once all of this state is saved on the stack, we can switch the stack pointer to the stack of a different task, and then restore all of the new task's state. We will save the stack pointer of the paused program at memory address 0x8001. Note the "pushad" instruction, which pushes all eight general-purpose registers onto the stack, and the "pushfd" instruction which saves the flags register.

```
SwitchTasks:
pushfd                       ; save flags
push ds                      ; save segment register
pushad                       ; save general-purpose registers

mov ebx, esp                 ; put the current program's stack
                             ; pointer in ebx

inc byte [es:0x8000]         ; increase the counter
cmp byte [es:0x8000], 90     ; see if it is time to switch
jnz DoneSwitch               ; if not, skip to the end

; Switch!
mov byte [es:0x8000], 0      ; reset the counter

mov dword ebx, [es:0x8001]   ; put the other program's stack
                             ; pointer into ebx

mov [es:0x8001], esp         ; save the current program's
                             ; stack pointer in memory

DoneSwitch:
mov esp, ebx                 ; switch to whichever stack pointer is in ebx

popad                        ; restore registers from current stack
pop ds                       ; restore data segment from current stack
popfd                        ; restore flags from current stack
ret                          ; resume execution
```

You may have noticed that the memory accesses are relative to the segment register ES, instead of the default DS. This is necessary to actually access memory locations 0x8000 and 0x8001. Recall that Program A and Program B each have memory accesses redirected to their own segment of memory, via the segment register DS. The segment selected in DS (for either program) is unable to access physical memory locations 0x8000 and 0x8001; it is restricted to a small area of the video memory.

There is one final step to make all of this work. The SwitchTasks function assumes that the stack pointer for another task is at memory location 0x8001, and that the stack to which it points contains a saved CPU state that can be loaded. This is true almost all of the time, as SwitchTasks saves the CPU state on the current program's stack before switching to another program. The one time that this is not true is when the system is first started.

When the system starts, we will start running program A. Eventually, Program A will have run for long enough, and it will be time for Program B to run. Program A calls SwitchTasks, which saves the CPU state for Program A to the stack for Program A. SwitchTasks then reads from memory address 0x8001 to find the stack pointer of Program B, and reads CPU state for Program B from that location. For this to work the first time, we will need to manually store a valid CPU state for Program B, so the first time SwitchTasks is called there will be a CPU state to restore. When Program B calls SwitchTasks, there will be a valid stack pointer at 0x8001 and a valid saved CPU state, both of which were saved when Program A was paused. Setting up a valid stack requires the items that are normally pushed by a previous call to SwitchTasks: the general-purpose registers, the segment register, the flags register, and the return address from the "call" instruction.

```
; Setup Program B's stack
mov esp, 0xB00          ; Program B's stack starts at 0xB00
push ProgramB           ; Return address, normally pushed by
                        ; "call SwitchTasks"
pushfd                  ; Flags, normally saved by SwitchTasks
push 0x20               ; DS, normally saved by SwitchTasks
pushad                  ; General purpose-registers, normally saved
                        ; by SwitchTasks

mov dword [0x8001], esp ; save ESP of paused program
```

Now when the first task switch reads from address 0x8001, it will obtain a stack pointer to the top of Program B's state. SwitchTasks will then restore the saved CPU state (general-purpose registers, segment register, and flags register), and then issue a "ret" to resume execution at the return address. You can see that Program B's data segment is used as the saved value for the segment register DS.

Bringing it all together, we can now setup an initial saved state for Program B, then start Program A. The two programs will then periodically take turns running for a short period of time, via the SwitchTasks function.

```
[ORG 0x7C00]

[BITS 16]

; Disable interrupts
cli

; Enter color mode
mov ax, 3
int 0x10

; Enable A20
call WaitForKeyboardController
mov al,0xD1
out 0x64,al
call WaitForKeyboardController
mov al,0xDF
out 0x60,al

; Load GDT
mov ax, 0x0000
mov ds, ax
lgdt [GDTPointer]

; Enable protected mode
mov eax, cr0
or  al, 1
mov cr0, eax

; Enter 32-bit protected mode
jmp 0x0010:ProtectedMode

WaitForKeyboardController:
mov cx, 1000                     ; Set retry count
CheckKeyboardController:
in al, 0x64                      ; Read from the keybaord controller
and al, 0x2                      ; See if bit 1 is set
loopnz CheckKeyboardController   ; If not, and the counter is above
                                 ; zero, try again
ret

[BITS 32]
ProtectedMode:

;Set default state
mov ax, 0x0008
mov ds, ax
mov es, ax
mov ss, ax

mov byte [0x8000], 0             ; Set SwitchTasks counter to zero

; Setup Program B's stack
mov esp, 0xB00                   ; Program B's stack starts at 0xB00
push ProgramB                    ; Return address for "ret" in SwitchTasks
pushfd                           ; Flags, normally saved by SwitchTasks
push 0x20                        ; DS, normally saved by SwitchTasks
```

```
pushad                      ; General purpose-registers, normally saved
                            ; by SwitchTasks

mov dword [0x8001], esp     ; save ESP of paused program

; Prepare A to run
mov esp, 0xA00              ; Program A's stack starts at 0xA00
mov eax, 0x18              ; Program A's memory segment is
mov ds, eax                ; GDT entry number 0x18
jmp ProgramA               ; Start running Program A


SwitchTasks:
pushfd                      ; save flags
push ds                     ; save segment register
pushad                      ; save general-purpose registers

mov ebx, esp                ; put the current program's stack
                            ; pointer in ebx

inc byte [es:0x8000]        ; increase the counter
cmp byte [es:0x8000], 90    ; see if it is time to switch
jnz DoneSwitch              ; if not, skip to the end

; Switch!
mov byte [es:0x8000], 0     ; reset the counter

mov dword ebx, [es:0x8001]  ; put the other program's stack
                            ; pointer into ebx

mov [es:0x8001], esp        ; save the current program's
                            ; stack pointer in memory

DoneSwitch:
mov esp, ebx                ; switch to whichever stack pointer is in ebx

popad                       ; restore registers from current stack
pop ds                      ; restore data segment from current stack
popfd                       ; restore flags from current stack
ret                         ; resume execution

ProgramA:
mov eax, 0                  ; Start writing at address 0x00000000
mov ecx, 0x3C0             ; Write 0x3C0 characters
LoopA:
mov byte [eax], 'A'         ; Set the character to 'A'
inc eax                     ; Move to the color byte
inc byte [eax]              ; Change the color
and byte [eax], 0x7F        ; Remove the blink bit
inc eax                     ; Move to the next character
loop LoopA, ecx             ; Keep going if not done
call SwitchTasks            ; Switch to another task
jmp ProgramA                ; Keep running forever

ProgramB:
mov eax, 0                  ; Start writing at address 0x00000000
mov ecx, 0x3C0             ; Write 0x3C0 characters
LoopB:
mov byte [eax], 'B'         ; Set the character to 'A'
inc eax                     ; Move to the color byte
inc byte [eax]              ; Change the color
and byte [eax], 0x7F        ; Remove the blink bit
inc eax                     ; Move to the next character
```

193

```
loop LoopB, ecx          ; Keep going if not done
call SwitchTasks         ; Switch to another task
jmp ProgramB             ; Keep running forever


GDTPointer:
dw 5 * 8 - 1
dd GDT

GDT:
dq 0x0000000000000000  ; 0x0000 - Null
dq 0x00CF92000000FFFF  ; 0x0008 - OS Data
dq 0x00CF9A000000FFFF  ; 0x0010 - Code
dq 0x0040920B80A0077F  ; 0x0018 - Program A Data
dq 0x0040920B8820077F  ; 0x0020 - Program B Data

; Pad to 512 bytes
TIMES 510-($-$$) DB 0
DW 0xAA55
```

12.5 Interrupts

The system is now capable of running two programs. Unfortunately, each program must signal when it is time to give other programs a chance to run. If any program gets stuck indefinitely, or creates an error, the whole system will halt; if the program is stuck, it will never call SwitchTasks. A more robust system can switch between tasks without depending on programs explicitly yielding control. Instead, the operating system must take control, and then decide if it is time to switch tasks. Fortunately, one of the many interrupts available to the CPU is a periodic timer.

The regular program flow can be interrupted (and redirected to a handler) for a variety reasons: invalid memory requests, dividing by zero, and signals on external wires are just several of the many possibilities. We will use an external signal that originates from a timer to periodically halt the currently the running program, and take control of the CPU via the interrupt handler.

Recall that a regular function call (using a "call" instruction) diverges from the regular flow of execution and jumps to the specified function. To resume execution from where the function was called, the "call" instruction pushes the current instruction pointer address – the value in EIP – onto the stack. At the end of the function, the "ret" instruction moves this value from the stack back into the EIP register, causing the program to resume from where it left off. Interrupts work in a very similar fashion.

The code for interrupt handlers is often in a different memory segment than the currently running program. The handler specifies in which code segment it resides, indicating to the CPU which value should be placed in the CS register before jumping to the handler. To resume the interrupted program, the original CS value will have to be restored, as well as the original instruction pointer (EIP). Hence, when an interrupt occurs, the CPU pushes the current instruction pointer address (EIP) and code segment value (CS) onto the stack, as well as the current EFLAGS value. At the end of a regular function, we use a "ret" instruction to restore the previous EIP value. At the end of an interrupt handler we can use an "iret" instruction to restore the EIP, CS, and EFLAGS values from the stack. This restores the interrupted program to its original state.

For the sake of example, we will handle two interrupts: general protection fault, and the system timer. A general protection fault is triggered when a running program attempts to access memory outside of its allowable range. This allows the operating system to take action (such as terminating the program) to remedy the situation. The timer interrupt is triggered whenever a specified amount of time has elapsed. The timer can be reconfigured to use different durations, but upon system startup it is left at 18 Hz by default (that is, it creates a signal eighteen times per second).

The handler for general protection faults will simply place an exclamation point in the top-left corner of the screen, then pause execution indefinitely, never returning.

```
ErrorHandler:
cli                                 ; ignore any other interrupts
mov word [es:0xB8000], 0xCF21 ; display red and white "!"
Hang:                               ; wait forever
jmp Hang
```

The timer handler is the same as the manual task-switching function – SwitchTasks – with two small exceptions. The first change is how we exit the interrupt handler. The SwitchTasks function simply used a "ret" instruction to restore the EIP value pushed by the "call" instruction; the interrupt handler must use an "iret" instruction to restore the three values pushed when the CPU detected the interrupt. In addition, the programmable interrupt controller must be told that the interrupt has been handled. This is accomplished by sending an acknowledgement using the "out" instruction.

The second change is much simpler. Because the CPU saves the value of the flags register before the interrupt handler begins, we no longer need to explicitly save it. The CPU will push the flags value onto the stack before the handler starts executing, and the "iret" instruction will restore the value while returning. This allows us to remove the "pushfd" and "popfd" instructions that manually saved the flags register.

```
TimerHandler:
push ds                       ; save segment register
pushad                        ; save general-purpose registers

mov ebx, esp                  ; put the current program's stack
                              ; pointer in ebx

inc byte [es:0x8000]          ; increase the counter
cmp byte [es:0x8000], 90      ; see if it is time to switch
jnz DoneSwitch                ; if not, skip to the end

; Switch!
mov byte [es:0x8000], 0       ; reset the counter

mov dword ebx, [es:0x8001]    ; put the other program's stack
                              ; pointer into ebx

mov [es:0x8001], esp          ; save the current program's
                              ; stack pointer in memory

DoneSwitch:
mov al,0x20
out 0x20,al                   ; tell the PIC the interrupt has been handled

mov esp, ebx                  ; switch to whichever stack pointer is in ebx

popad                         ; restore registers from current stack
pop ds                        ; restore segment register
iret                          ; resume execution
```

We now need to instruct the CPU to call these handlers in response to the appropriate interrupts. The CPU uses a table of handler entries to determine what function should be called in the event of a specific interrupt. Some interrupts are triggered internally, while others come from external sources. The first 32 (0x20) handlers are reserved for internal events, such as an application dividing by zero.

The first step in accepting external interrupts is to indicate which range of interrupt table entries should be dedicated to handling the 16 external interrupts. This is accomplished by sending the desired range to the programmable interrupt controller, or PIC.

```
; Map external interrupts 0-15 to entries 0x20-0x2F
mov al,0x15
out 0x20,al
out 0xA0,al

mov al,0x20
out 0x21,al
add al,8
out 0xA1,al

mov al,4
out 0x21,al
mov al,2
out 0xA1,al

mov al,0x01
```

```
out 0x21,al
out 0xA1,al
```

The specific purpose of each bit is detailed in the specification for the programmable interrupt controller chip, and is beyond the scope of this book. The important fact is that the code shown above directs the 16 external interrupts to interrupt table entries numbered 0x20 through 0x2F. When external interrupt number 1 occurs (which is a keyboard event), the CPU will look up entry number 0x21 in the interrupt table to find the handler that should be called.

Now that we have indicated which entries correspond to external interrupts, we can begin creating the interrupt table entries themselves. Each table entry consists of eight bytes. Each block in the following diagram represents sixteen bits (two bytes).

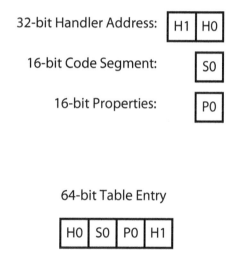

The CPU calls the appropriate handler when an interrupt is triggered. The pointer to the current instruction (the value in register EIP) is always relative to the currently selected code segment (indicated by the value in register CS). The interrupt handling code is likely within a different code segment than the currently running application's code, so the table entry includes a segment number as well as the address of the handler function. We will always keep the properties bits set to 0x8E00, which indicates that the handler is present.

We will construct the table at memory location 0x1000000. General protection fault is internal interrupt number thirteen. Because each entry is eight bytes, entry number thirteen (0xD) begins at address 0x1000068 (0x1000000 + 0xD × 0x8). We mapped the sixteen external interrupts to handlers numbered 0x20 – 0x2F. The timer is external interrupt number zero, so it uses handler number 0x20 at address 0x1000100. Using the format illustrated in the previous diagram, we can now create both entries in the table.

```
; Create general protection fault handler entry
mov eax, 0x100068        ; Entry number 13 (0x0D)
mov word [eax], ErrorHandler  ; Set the handler
add eax, 0x2
mov word [eax], 0x0010   ; Set the code segment
add eax, 0x2
mov word [eax], 0x8E00   ; Set the properties
add eax, 0x2
mov word [eax], 0x0000   ; High half of handler

; Create timer handler entry
mov eax, 0x100100        ; Entry number 32 (0x20)
mov word [eax], TimerHandler  ; Set the handler
add eax, 0x2
mov word [eax], 0x0010   ; Set the code page
add eax, 0x2
mov word [eax], 0x8E00   ; Set the properties
add eax, 0x2
mov word [eax], 0x0000   ; High half of handler
```

Similar to the GDT, the interrupt descriptor table (or "IDT") is loaded with a special instruction that accepts the location and size of the table. The location and size are specified in the same manner as the GDT, with a value that is one less than the size of the entire table, followed by the address of the table.

```
IDTPointer:
dw 0x21 * 8 - 1
dd 0x100000
```

The table is now ready to be loaded.

```
lidt [IDTPointer]
```

The CPU can now look up the appropriate handler for the two interrupts we have chosen to handle. All that remains to be done is to enable the interrupts. The external programmable interrupt controller uses a set of sixteen bits to indicate which of the sixteen external interrupts should be ignored. The only external interrupt with a valid handler is the timer – external interrupt number zero – so only one bit number zero will be cleared. The other remaining fifteen bits are set to one, so all other external interrupts will be ignored.

```
mov al, 11111110b        ; Ignore everything except the timer
out 0x21, al
mov al, 11111111b
out 0xA1, al
```

The final step is to enable interrupt handling on the CPU. The first instruction of each program so far has been the "cli" instruction, which disables interrupts. We can now issue the "sti" instruction to enable them.

```
sti          ; Enable interrupts
```

The timer will now trigger an interrupt approximately eighteen times per second. The timer handler switches tasks every ninety times it runs, resulting in a task switch every five seconds. If you run the code in the emulator it may take significantly longer than five seconds, as the emulator may not be able to keep up. Incorporating all of these changes yields the following code.

```
[ORG 0x7C00]

[BITS 16]

; Disable interrupts
cli

; Enter color mode
mov ax, 3
int 0x10

; Enable A20
call WaitForKeyboardController
mov al, 0xD1
out 0x64, al
call WaitForKeyboardController
mov al, 0xDF
out 0x60, al

; Load GDT
mov ax, 0x0000
mov ds, ax
lgdt [GDTPointer]

; Enable protected mode
mov eax, cr0
or  al, 1
mov cr0, eax

; Enter 32-bit protected mode
jmp 0x0010:ProtectedMode

WaitForKeyboardController:
mov cx, 1000                     ; Set retry count
CheckKeyboardController:
in al, 0x64                      ; Read from the keybaord controller
and al, 0x2                      ; See if bit 1 is set
loopnz CheckKeyboardController   ; If not, and the counter is above
                                 ; zero, try again
ret

[BITS 32]
ProtectedMode:
```

```
;Set default state
mov ax, 0x0008
mov ds, ax
mov es, ax
mov ss, ax

mov byte [0x8000], 0            ; Set switch counter to zero


mov al,0x15                     ; Map external interrupts
out 0x20,al                     ; 0-15 to entries 0x20-0x2F
out 0xA0,al

mov al,0x20
out 0x21,al
add al,8
out 0xA1,al

mov al,4
out 0x21,al
mov al,2
out 0xA1,al

mov al,0x01
out 0x21,al
out 0xA1,al

; Create general protection fault handler entry
mov eax, 0x100068              ; Entry number 13 (0x0D)
mov word [eax], ErrorHandler   ; Set the handler
add eax, 0x2
mov word [eax], 0x0010         ; Set the code segment
add eax, 0x2
mov word [eax], 0x8E00         ; Set the properties
add eax, 0x2
mov word [eax], 0x0000         ; High half of handler

; Create timer handler entry
mov eax, 0x100100             ; Entry number 32 (0x20)
mov word [eax], TimerHandler  ; Set the handler
add eax, 0x2
mov word [eax], 0x0010        ; Set the code page
add eax, 0x2
mov word [eax], 0x8E00        ; Set the properties
add eax, 0x2
mov word [eax], 0x0000        ; High half of handler

lidt [IDTPointer]

mov al, 11111110b            ; Ignore everything except the timer
out 0x21, al
mov al, 11111111b
out 0xA1, al

; Setup Program B's stack
mov esp, 0xB00              ; Program B's stack starts at 0xB00
push 0x20                   ; DS, normally saved by SwitchTasks
pushfd                      ; Flags, CS (code segment), and
push cs                     ; EIP (return address), for "iret", normally
push ProgramB               ; pushed by the CPU when an interrupt occurs.
push 0x20                   ; DS, normally saved by TimerHandler
pushad                      ; General purpose-registers, normally pushed
                            ; by TimerHandler
```

```
mov dword [0x8001], esp      ; Save ESP of paused program for TimerHandler

; Prepare A to run
mov esp, 0xA00               ; Program A's stack starts at 0xA00
mov eax, 0x18                ; Program A's memory segment is
mov ds, eax                  ; GDT entry number 0x18

sti                          ; Enable interrupts
jmp ProgramA                 ; Start running Program A


ErrorHandler:
cli                          ; ignore any other interrupts
mov word [es:0xB8000], 0xCF21 ; display red and white "!"
Hang:                        ; wait forever
jmp Hang

TimerHandler:
push ds                      ; save segment register
pushad                       ; save general-purpose registers

mov ebx, esp                 ; put the current program's stack
                             ; pointer in ebx

inc byte [es:0x8000]         ; increase the counter
cmp byte [es:0x8000], 90     ; see if it is time to switch
jnz DoneSwitch               ; if not, skip to the end

; Switch!
mov byte [es:0x8000], 0      ; reset the counter

mov dword ebx, [es:0x8001]   ; put the other program's stack
                             ; pointer into ebx

mov [es:0x8001], esp         ; save the current program's
                             ; stack pointer in memory

DoneSwitch:
mov al,0x20
out 0x20,al                  ; tell the PIC the interrupt has been handled

mov esp, ebx                 ; switch to whichever stack pointer is in ebx

popad                        ; restore registers from current stack
pop ds                       ; restore segment register
iret                         ; resume execution


ProgramA:
mov eax, 0                   ; Start writing at address 0x00000000
mov ecx, 0x3C0               ; Write 0x3C0 characters
LoopA:
mov byte [eax], 'A'          ; Set the character to 'A'
inc eax                      ; Move to the color byte
inc byte [eax]               ; Change the color
and byte [eax], 0x7F         ; Remove the blink bit
inc eax                      ; Move to the next character
loop LoopA, ecx              ; Keep going if not done
jmp ProgramA                 ; Keep running forever

ProgramB:
mov eax, 0                   ; Start writing at address 0x00000000
```

```
mov ecx, 0x3C0              ; Write 0x3C0 characters
LoopB:
mov byte [eax], 'B'        ; Set the character to 'A'
inc eax                    ; Move to the color byte
inc byte [eax]             ; Change the color
and byte [eax], 0x7F       ; Remove the blink bit
inc eax                    ; Move to the next character
loop LoopB, ecx            ; Keep going if not done
jmp ProgramB               ; Keep running forever


GDTPointer:
dw 5 * 8 - 1
dd GDT

GDT:
dq 0x0000000000000000 ; 0x0000 - Null
dq 0x00CF92000000FFFF ; 0x0008 - OS Data
dq 0x00CF9A000000FFFF ; 0x0010 - Code
dq 0x0040920B80A0077F ; 0x0018 - Program A Data
dq 0x0040920B8820077F ; 0x0020 - Program B Data

IDTPointer:
dw 0x21 * 8 - 1
dd 0x100000

; Pad to 512 bytes
TIMES 510-($-$$) DB 0
DW 0xAA55
```

Each of Program A and Program B are given enough space in memory to write 0x3C0 colored characters. You can try changing the counter from 0x3C0 to 0x3C1 to observe the general protection fault handler interrupting programs that access memory outside of their assigned range.

This simple system demonstrates how CPU features can be used to automatically switch between tasks and catch accesses to memory outside of a program's designated range. There are still many ways for these programs to gain access to memory outside of their default data segment (the simplest probably being to just explicitly use a different data segment), as well as perform other dangerous operations. A more complete operating system can use built in security features to limit the operations that programs can perform.

12.6 Exercises

1. Program A and Program B are given equal amounts of time to run. How could one program be allowed to run for a larger portion of time than the other?

2. Modern operating systems allow many more than two programs to be run at once. Does switching between running programs correspond to task switching in the sample operatring system?

3. Could the sample operating system support multiple instances of the same program? For example, could two copies of Program A be run at the same time, instead of one Program A and one Program B?

Part III: Software Programming

13

C# Basics

13.1 The Environment

The machine-code programming shown in the previous section is capable of creating virtually any program you desire. Using an operating system to switch tasks allows many programs to run concurrently and independently. Unfortunately, writing programs entirely in assembly presents several problems.

The most obvious issue with assembly programming is the amount of time it takes to create programs. Reusing common sections of code can lessen the burden, but ultimately a new way of creating programs is desirable. Rather than specifying each instruction for the CPU to execute, we can use a different language to describe the desired behavior at a higher level. A specialized program – a compiler – can then generate CPU instructions that exhibit the specified behavior.

The second problem that arises when writing each instruction is a lack of compatibility. As noted earlier, different CPUs support different modes: 64-bit, 32-bit, etc. Even within each mode, some CPUs offer extra extensions to the instruction set for special-purpose computation. When writing a program in assembly, you must choose which instructions you use. 64-bit instructions cannot be carried out on a 32-bit CPU, and many special-purpose instructions are not available on processors from other manufacturers. Manually entering each instruction forces you to choose a subset of processors on which your program will run.

Using C#, or one of several other similar languages, resolves both of these issues. C# is a higher-level language, meaning it describes more complex operations. These operations can then eventually be translated into CPU instructions.

The second problem – compatibility – is addressed by performing a two-stage translation. The code is written in a high-level form, describing the behavior of the program. The programmer then uses a compiler to translate this into intermediate instructions. These intermediate instructions are similar to CPU instructions, but in a generic form that is not specific to any one CPU. The program is sent to the user in the intermediate form. When the user runs the program, the generic instructions are mapped to the specific instructions that the user's CPU can execute. This allows the same program to run on a wide variety of computers, but also requires each user to have some software installed to perform this final instruction mapping. This extra software is included in a package referred to as a "runtime library", or simply "runtime", and is often already installed on new computers.

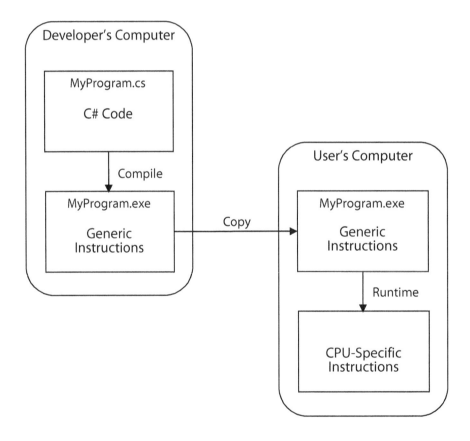

13.2 Writing and Compiling Code

The following examples use the free tools from Microsoft® to compose the C# code and compile it into a program. Once you have installed Microsoft Visual C#®, you can create a new project by selecting "New Project" from the "File" menu. Choose the "Console Application" type of project, and give it any name you please. You should be presented with a simple program that you can begin expanding. Modify the code to match the following program.

```csharp
using System;

namespace Application
{
    class Program
    {
        static void Main(string[] args)
        {
            // This is a comment

            /* So
               is
               this */

            Console.WriteLine("Hello!");
            Console.ReadKey();
        }
    }
}
```

Pressing the green play button (or selecting "Start Debugging" from the "Debug" menu) will run the program. We will examine the program from the inside out. The innermost lines – the ones beginning with "Console." – are the operations that are carried out. The first operation writes "Hello!" to the console, while the second operation waits for a key to be pressed.

C# is somewhat similar to the assembly language that was used in the previous section. Some lines are used for organizational purposes or compiler directives, while other lines actually generate code to be executed. In this case, only the two lines beginning with "Console" are carried out; the other lines direct the compiler to compile the program in a certain way. You can try adding similar "WriteLine" operations to print more words on the screen. Take note that each line that is actually carried out ends in a semicolon, while the other directives for the compiler do not. It is the semicolon that marks the end of an operation; extra spaces or extra lines do not affect the statement. In addition, any text between a "//" and the end of a line is a comment, as is text between "/*" and "*/".

When you installed the programming tools to use C#, both the compiler and runtime were installed. Choosing "Start Debugging" from within Visual C# compiles the program into the generic instructions, and then runs the program using the runtime. You could send this program to anyone else with the runtime installed and the program would be automatically translated to instructions most appropriate for their CPU. Your program can run on any platform that has a runtime available, including computers, cellular phones, PDAs, webpages, and video game consoles. Selecting "Save All" from the "File" menu will allow you to save your project to a specified location. You can then find your program (e.g. "program.exe") in a sub-folder of the location you specified. Feel free to run the program by itself, or send it to a friend.

13.3 Variables

One of the most basic operations that a program can carry out is to store and retrieve data. In the assembly examples, various tables (GDT, etc.) and state items (the currently running task) were stored in pre-determined memory locations. Although quite functional, manually selecting locations for each set of data can become quite tedious; whenever a new item is to be added, you must consider where there is enough free memory for it to be stored.

Variables allow you to simply specify what type of data you would like to store, and then assign a name to that data. The placement of the data in memory will be handled automatically, as will deciding when to move the data in and out of registers for manipulation. The first step in creating a variable is selecting an appropriate type. The type of the variable determines its size and how it should be manipulated. For instance, when the values in two variables are added together, the compiler needs to know whether to treat the numbers as signed (possibly negative, in two's-complement notation) or unsigned (never negative).

The basic types fall into four categories: integers (e.g. 5, 37, or 184), floating-point numbers (e.g. 5.7, 3.14, or 123.45678), Boolean values (true or false), and characters (e.g. 'a', 'C', or '!'). Several different sizes are available. If you wanted to store the result of "60 × 1000", you could not use a signed 8-bit integer (which can express -128 to 127, or -2^7 to $2^7 - 1$), or an unsigned 8-bit integer (which can express 0 to 255, or 2^0 to $2^8 - 1$). You could not use a signed 16-bit value (-2^{15} to $2^{15} - 1$), but you could use an unsigned 16-bit value (0 to 2^{16}). The size of the values you wish you use will determine which type is most appropriate.

Type	Content	Values
sbyte	8-bit signed integer	-2^7 to 2^7-1
short	16-bit signed integer	-2^{15} to $2^{15}-1$
int	32-bit signed integer	-2^{31} to $2^{31}-1$
long	64-bit signed integer	-2^{63} to $2^{63}-1$
byte	8-bit unsigned integer	0 to 2^8-1
ushort	16-bit unsigned integer	0 to $2^{16}-1$
uint	32-bit unsigned integer	0 to $2^{32}-1$
ulong	64-bit unsigned integer	0 to $2^{64}-1$
float	32-bit floating point	7 significant digits
double	64-bit floating point	16 significant digits
decimal	128-bit floating point	28 significant digits
bool	Truth value	true or false
char	Character	Letters, symbols, numerals, etc.

The second part of creating a variable is assigning a name. Names are case-sensitive; "x" is a different variable than "X".

```
int x;
```

Once the variable has a type and a name, you can begin to use it. Values can be placed into a variable using the assignment operation, which is indicated by a single equals sign. This can be done when the variable is created, or anytime after. The following code (eventually) sets both "first" and "second" to the value six.

```
int first;
int second = 7;
first = 6;
second = first;
```

The type of a variable determines what sort of data the variable holds; an "int" stores an integer, a char stores a character, and so on. There are two main kinds of types: value types, and reference types.

The value type is the simplest. When you create a variable that is a value type, the variable will hold the actual values you want to use. All of the types listed in the previous table are value types. When you create a variable of the "int" type, a four-byte memory location will be chosen to store a 32-bit integer. When you assign a value to that variable, the value is simply written to the selected memory location.

Examining the previous code sample in more detail can help illustrate the behavior of value types. The first line reserves four bytes to store a 32-bit integer. The second line reserves another four bytes, and writes the value 0x00000007 into them. The third line writes the four-byte value 0x00000006 into the first four-byte memory area. The fourth line reads the value from the first four-byte memory area (0x00000006), and then copies it into the second four-byte memory area.

The reference type is slightly more complicated. Instead of storing the actual data, a reference type stores the address of some data. When a reference type variable is created, it initially refers to nowhere, or "null". Null is a special value used to signify that a reference type does not contain a valid memory address. Reference types will be used in the following sections in more detail.

C# also provides the basic binary and arithmetic operations that were present in various assembly instructions. Integer variables can be combined using any of the following operators.

Symbol	Operation	Result if a = 0x0D and b = 0x03
a + b	Add a and b	0x10
a++	Increase a	0x0E
a--	Decrease a	0x0C
a - b	Subtract b from a	0x0A
a * b	Multiply a and b	0x27
a / b	Divide a by b	0x04
a & b	a AND b	0x01
a | b	a OR b	0x0F
a ^ b	a XOR b	0x0E
~a	Invert a	0xF2
a << b	Shift a left by b bits	0x68
a >> b	Shift a right by b bits	0x01

Several operations can be combined in the same line. Parenthesis can cause some parts of the expression to be evaluated before others.

```
int a;
int b = 2;
a = 2 * (b + 3);
```

The code shown above will place the value 10 into the variable "a". The part of the statement in the parenthesis is evaluated first, followed by the multiplication. You can observe this output by running the following program.

```
using System;

namespace Application
{
    class Program
    {
        static void Main(string[] args)
        {
            Console.WriteLine("Hello!");

            int a;
            int b = 2;
            a = 2 * (b + 3);
            Console.WriteLine(a);

            Console.ReadKey();
        }
    }
}
```

The final basic type is the "string". A string is simply a sequence of characters. Strings can be created by enclosing a sequence of characters in double quotes (while a single character, of the "char" type, can be created by enclosing a character in single quotes). The "+" operator allows values to be added to an existing string (after being converted to a string). For example, when the following code adds an integer to a string, a new string is created that contains the value of the original string with a string representation of the integer's value appended to the end.

```
int myInteger = 10;
string myString = "hello " + myInteger;
Console.WriteLine(myString);
```

The sequence of characters "hello ", enclosed in double quotes, creates a new string containing the word "hello" followed by a space. Adding the integer variable to end creates a new string: "hello 10". A reference to this new string is stored in the variable "myString". The string is then printed to the console. The same result could be achieved with a single line of code:

```
Console.WriteLine("hello " + 10);
```

or

```
Console.WriteLine("hello 10");
```

In addition to the type to use, variables creation can also specify several modifiers. A "const" variable is a fixed value that cannot be changed by the program. Constant, or "const", variables are typically used to give a more meaningful name to a numerical value.

Not only is this name easier to read, but also much more convenient to change in the future. Using a "const" variable instead of the value itself means only the variable creation would need to be changed, instead of each place where the value is used.

```
const int BitsPerByte = 8;
const double Pi = 3.14;
const char CurrencySymbol = '$';
```

Although the number of bits in a byte is probably constant, it is foreseeable that you may want to add more digits to pi for more precision, or change your currency symbol. In both of these cases, only the variable would need to be change, and every line of code that used the variable would automatically be affected, rather than searching your code for each place you used "3.14" or a dollar sign.

Another modifier that can be added is the "static" specification. Static items are only created once per running copy of an application. Although many parts of the application may use them, a single copy is shared between all parts of the program. They will be covered in more detail in the following section.

13.4 Arrays

If you are working with a large number of related variables, creating names for each variable is far too tedious to be worthwhile. Instead, a large group of variables can be created with a name assigned to the group, rather than each variable. This group is called an "array". An array is simply a group of variables of the same type.

Arrays are reference types. The array variable holds the address of the group of data, and initially contains null. An array variable is created by following an existing type with square brackets. The following code creates a variable that can hold the address of a group of integers.

```
int[] myArray;
```

Of course, we have not created a group of integers yet, only a variable that can hold an address. The next step is to actually create the group of integers, storing its address in our variable. The following line creates a group of ten integers.

```
myArray = new int[10];
```

With a valid address stored in the variable, we can now access each of the ten integers at that address. The ten integers are accessed by "index", using numbers zero through nine.

```
myArray[0] = 5;          // set the first integer to a value of 5
myArray[9] = 6;          // set the last integer to a value of 6
```

With the basics of array use out of the way, we can now explore the significance of arrays being a reference type, rather than a value type. Examine the following program.

```
using System;

namespace Application
{
    class Program
    {
        static void Main(string[] args)
        {
            Console.WriteLine("Hello!");

            int[] array1 = new int[10];
            int[] array2 = array1;

            array1[0] = 3;
            array2[0] = 6;

            Console.WriteLine(array1[0]);

            Console.ReadKey();
        }
    }
}
```

After printing "Hello!", the program creates two array variables and stores the values three and six. Finally, the program prints the value contained in the first integer of the first array. Because these are reference types, the number six is printed.

You can see that when array2 is created, it is assigned the value in array1. Array variables are reference type variables, so they only contain an address. The first array – array1 – stores the address of a new block of ten integers. The second array – array2 – stores the address that is contained in array1. Hence, both variables contain the address of the same block of integers. Knowing this, it is clear to see that when the first element of array2 is set to a value of six, it overwrites the value of three that was placed there on the previous line. Although array1 and array2 are two separate variables, they both refer to the same group of integers.

13.5 Conditionals

So far, all of the code we have written has executed line by line, from top to bottom. The assembly part of this book demonstrated several ways to jump around to various parts of a program when certain conditions were met (e.g. a register containing zero). C# contains several constructs that will generate these jump instructions. Conditional statements carry out a section of code if a truth value is equal to "true". The first step to using conditional expressions is to understand how truth (or "Boolean") values work in C#.

The variable type "bool" can only contain one of two values: true or false.

```
bool first = true;
bool second = false;
```

Several comparison operators are available to generate "true" or "false" based on the values of existing variables.

Operator	Result
a == b	True only if a is equal to b
a != b	True only if a is not equal to b
a > b	True only if a is greater than b
a >= b	True only if a is greater than b, or equal to b
a < b	True only if a is less than b
a <= b	True only if a is less than b, or equal to b

Various logical operators are available to combine Boolean values as well.

Operator	Result
a && b	True if a is true and b is true
a || b	True if a is true or b is true
a ^ b	True if exactly one of a and b are true
a & b	True if a is true and b is true
a | b	True if a is true or b is true

The first two operators are special "short-circuit" operators. They yield the same result as the normal "and" and "or" operators, but only evaluate the second parameter ("b", in the table) if it could change the outcome. For example, if "a" is true then "a || b" is true regardless of the value of b. In this situation, the value of b would not be retrieved/calculated.

```
int a = 6;
int b = 10;
bool c = (a > b) || (a > 4);          // sets c to "true"
bool d = b < a;                        // sets d to "false"
```

The first, and simplest conditional operation is the "if" statement. The if statement carries out a block of code if the specified expression evaluates to true. If the expression does not evaluate to true, the block of code is skipped.

```
int a = 5;

if (a < 10)
{
    Console.WriteLine("a is less than 10");
}

if(a > 10)
{
    Console.WriteLine("a is greater than 10");
}

if (a == 10)
{
    Console.WriteLine("a is equal to 10");
}
```

The preceding code will print "a is less than 10", and nothing else. Related if statements can be grouped together using "else if" and "else". Only one of the blocks of code in an if/else-if/else group will be carried out. Whichever block is the first to have its condition met will be run, and all others will be skipped. If none of the conditions match, the "else" block of code will be run (if there is one). Note that there are no semicolons at the end of the "if" statement, or the braces enclosing the blocks of code.

The first statement must be an "if" statement. The "if" statement can be followed by any number of "else if" statements (including zero), which can be followed by at most one "else" statement.

```
int a = 5;
int b = 6;

if(a == 10)
{
    Console.WriteLine("a is ten");
}
else
{
    Console.WriteLine("a is not ten");
}

if (a + b == 10)
{
    Console.WriteLine("a plus b is ten");
}
else if(a + b == 11)
{
    Console.WriteLine("a plus b is eleven");
}
else
{
    Console.WriteLine("a plus b is not ten or eleven");
}
```

The "switch" statement is often more convenient for longer lists of comparisons. "Switch" statements are similar to "if" statements, but are slightly more restrictive. The "switch" statement will match a specified value to one of the listed cases. If none of the cases match the specified value, the default case is carried out, if specified. Switch statements cannot use variables as cases; only fixed (constant) values are allowed. Also note that each block of code for a case is followed by a "break;". The following code will print "The sum is eleven", and nothing else.

```
int a = 5;
int b = 6;

switch (a + b)
{
    case 10:
    {
        Console.WriteLine("The sum is ten");
    }
    break;

    case 11:
    {
        Console.WriteLine("The sum is eleven");
    }
    break;

    default:
    {
        Console.WriteLine("The sum is not ten or eleven");
    }
    break;
}
```

Instead of just selecting a block of code to execute, we can also create a program that repeats a block of code. The easiest way to repeat a block of code is using the "while" statement. The "while" statement repeats a block of code while a given expression resolves to true. The expression is checked each time before the block of code is run, so if the expression is never true then the block will never be run.

```
int a = 5;
while (a > 0)
{
    Console.WriteLine("a is greater than zero");
    a--;
}
```

The preceding code will print "a is greater than zero" five times: once when a is equal to five, four, three, two, and one. The expression "a > 0" is false once a is decreased to zero, so the block is no longer repeated, and execution carries on after the block.

A very similar construct is the "do-while" loop. This loop is the same as a "while" loop, except for the first pass. A "do-while" loop always carries out the enclosed block of code once, then checks the expression before each subsequent loop. The following code will print "a is greater than zero" once, despite that fact that a is not greater than zero.

```
int a = 0;
do
{
   Console.WriteLine("a is greater than zero");
   a--;
} while (a > 0);
```

The pattern of creating a variable for counting and repeating a block of code is so common that a special construct is available to make it easier to write. The "for" loop integrates the creation of a counter variable, an expression to describe when to stop looping, and a statement to change the counter. Within the parenthesis following the "for" keyword is a statement, a Boolean expression, and another statement. The loop executes the first statement, then checks the Boolean expression. If the Boolean expression is true, the block of code is carried out, followed by the final statement in the for loop. If the Boolean expression resolves to false then execution carries on after the "for" loop.

```
for (int a = 5; a > 0; a--)
{
   Console.WriteLine("a is greater than zero");
}
```

The preceding code is equivalent to the following code.

```
int a = 5;
while (a > 0)
{
   Console.WriteLine("a is greater than zero");
   a--;
}
```

13.6 Debugging

You will inevitably need to do some debugging as you proceed through the basics of using C#. A common form of debugging during development is accomplished by setting "breakpoints" to examine the state of a program.

A breakpoint is a temporary setting that causes your program to halt when a specified line of code is reached. Breakpoints are only for debugging purposes and do not alter the file in which your program is stored (e.g. program.exe). Breakpoints are set by a debugger. In Visual C#, you can set a breakpoint by clicking the margin to the left of a line, or by pressing F9. You can then start your program (by pressing F5, or choosing "Start Debugging" from the menu) and make use of the breakpoint.

When the program reaches a breakpoint, execution is paused and you can examine many parts of the application. You can examine different sections of your program by using various windows available in the "Windows" submenu of the "Debug" menu. You can also hold your mouse cursor over variables to see their current value. Once you have discovered the values you desire, you can press F5 again (or select "Resume") to resume execution.

You can manually halt execution without breakpoints if the need should arise. If you started your program with the debugger, you can simply press the pause button (or Ctrl + Alt + Break) to suspend execution, wherever it may be. This is often useful if your program is stuck somewhere and you would like to see where.

13.7 Exercises

1. What is the maximum index that can be used with an array of fifty-seven items?

2. What is the type of the value within parenthesis in "if", "while", and "do-while" statements?

3. How can you test an integer to discover if bit number five is set to a one?

14

Program Structure

14.1 Classes

The "class" is the fundamental organizational unit in C#. A class is a grouping of data and functionality. Classes are first described (or "prototyped"), by listing the data and functionality they should contain. Once described, an instance of the class can be created. Creating an instance of (or "instantiating") a class reserves an area of memory that is sufficiently large to hold all of the data that the class requires. Many instances of a class can be created, and each will contain the variables specified in the class prototype.

Classes can be grouped together in "namespaces" for organizational purposes. All classes between the opening and closing brace following a "namespace" statement will belong to the specified namespace. Classes are identified by their namespace, followed by a period, followed by the name of the class. The code below creates the class "MyClass" within the namespace "MyNamespace". Other sections of code can refer to this class with the name "MyNamespace.MyClass".

```
using System;

namespace Application
{
    class Program
    {
        static void Main(string[] args)
        {
            Console.WriteLine("Hello!");
            Console.ReadKey();
        }
    }
}

namespace MyNamespace
{
    class MyClass
    {
    }
}
```

When referring to a class, the namespace can be omitted if the class is in the same namespace as the code in question, or a "using" statement at the top of the file specifies the namespace. The "Console" class in previous examples belongs to the "System" namespace. If the "using System;" line is not present at the top of the file, the output statements would have to be "System.Console.WriteLine(...);", instead of just "Console.WriteLine(...);".

Now that the new class has a name (and namespace) we can begin adding members to the class. Member variables are created in same manner as other variables, except they are placed inside braces that enclose the class, and outside any other braces.

```
using System;

namespace Application
{
    class Program
    {
        static void Main(string[] args)
        {
            Console.WriteLine("Hello!");

            MyNamespace.MyClass intstance;
            instance = new MyNamespace.MyClass();

            Console.ReadKey();
        }
    }
}

namespace MyNamespace
{
    class MyClass
    {
        int myInteger;
```

```
            char myCharacter;
        }
}
```

After printing "Hello!", a variable is created that can hold a reference to (the memory address of) an instance of the new class. The following line creates a new instance of the class in memory. Creating an instance of the class reserves a section of memory large enough to hold all of the variables that the class should contain. Static variables are an exception to this pattern. If a variable is made "static", one instance of that variable is created when the program starts and is shared between all instances of the class. Creating more instances of the class will not create more of the static variable, and any changes to the variable will affect all instances of the class (as they all share the same variable).

Each instance of the class now contains an integer and a character, but neither can be used from outside the class yet. By default, class member variables are "private", meaning that outside classes cannot access them. Simply making member variables "public" would allow any outside classes to manipulate them. To retain control over what happens to internal variables, classes often only allow variable manipulation through a "property". A property looks like a variable, but is actually a pair of functions. Whenever a value is assigned to a property, the "set" accessor is executed. Similarly, whenever the value of a property is read, the "get" accessor is carried out.

```
using System;

namespace Application
{
    class Program
    {
        static void Main(string[] args)
        {
            Console.WriteLine("Hello!");

            MyNamespace.MyClass instance;
            instance = new MyNamespace.MyClass();

            instance.MyCharacter = 'A';
            int integer = instance.MyInteger;

            Console.ReadKey();
        }
    }
}



namespace MyNamespace
{
    class MyClass
    {
        int myInteger;
```

```
        char myCharacter;

        public int MyInteger
        {
            get
            {
                Console.WriteLine("Reading myInteger");
                return myInteger;
            }
        }

        public char MyCharacter
        {
            get { return myCharacter; }
            set
            {
                if (value == 'A' || value == 'B')
                {
                    myCharacter = value;
                }
            }
        }
    }
}
```

You can see that MyClass now has two properties, one for each of the two member variables. The property that facilitates access to the character variable has both a "get" and "set" accessor, allowing external classes to both get and set the value. The get accessor returns the value that should be obtained when an outside class reads the property.

Set accessors are given a special variable named "value" that contains the value the outside class assigned to the property. The "set" accessor for the character property only allows the characters "A" or "B" to be written to the variable. Validating values being set is a common use of properties. Without the use of a property, outside classes could simply write any value into the variables, even if you only wanted either "A" or "B".

The integer property only contains a "get" accessor, meaning outside classes can only read the property, not write to it. In addition to returning a value to the outside class, the "get" accessor displays some text whenever the property is read. These properties are fairly simple, but any amount of code can be placed into an accessor.

14.2 Methods

Classes can contain functionality in addition to data (and its accessors). A method is a block of code that can be given several values, carry out a sequence of operations, and can return a single value. Methods can be called from other code, and can return a value back to the caller. This is the C# version of the "call" and "ret" instructions in assembly.

The "main" method is a special method that is automatically called by the runtime when the program starts. When you exit the main method, your program will exit. A method can exit explicitly with a "return;" statement, or automatically when execution reaches the closing brace of the method. If the method indicates that it is to return a value, the return statement must include a value of the specified type.

The "main" method also has a "static" modifier. Making a method "static" has an effect similar to making a variable "static". Normally, a method is called with respect to an instance of the class. When the method refers to a member variable of the class, it is directed to the variables in the specified instance of the class. Static methods run without an instance of the class, so an instance does not have to be created to call a static method. Consequently, a static method cannot access non-static member variables (it would not know which instance of the class to access, and an instance may not even exist). The "WriteLine" method in the "Console" class is static. You can see that we have been calling it without ever creating an instance of the "Console" class.

A method is created by specifying the return type, followed by the name of the method, followed by the list of arguments. The following code adds two methods, one of which is static.

```
using System;

namespace Application
{
    class Program
    {
        static void Main(string[] args)
        {
            Console.WriteLine("Hello!");

            MyNamespace.MyClass.StaticMethod(65);

            MyNamespace.MyClass instance;
            instance = new MyNamespace.MyClass(123);

            instance.MyCharacter = 'A';
            int integer = instance.MyInteger;

            Console.WriteLine("Is character B? " +
                instance.IsCharacterB());

            Console.ReadKey();
        }
    }
}

namespace MyNamespace
{
    class MyClass
    {
        int myInteger;
        char myCharacter;
```

227

```
public MyClass(int integerValue)
{
    myInteger = integerValue;
}

public int MyInteger
{
    get { return myInteger; }
}

public char MyCharacter
{
    get { return myCharacter; }
    set { myCharacter = value; }
}

public static void StaticMethod(int argument)
{
    string message = "Static method called!";
    message = message + " Argument: " + argument;
    Console.WriteLine(message);
}

public bool IsCharacterB()
{
    return (myCharacter == 'b' || myCharacter == 'B');
}
    }
}
```

The static method is called before any instances of the class are created and it does not use any class member variables. The "IsCharacterB" method cannot be static, as it needs to access the member variables of an instance (the "myCharacter" member).

Every class can have a special method that is called when the class is instantiated. If the class contains a method with the same name as the class (and no return type), it will be called whenever an instance of the class is created. This special method, called a "constructor", can also contain arguments, which are passed in by the code creating an instance of the class. The sample above uses the constructor to store a value in the integer member variable.

The main method will print "Hello!", followed by "Static method called! Argument: 65". The main method then creates an instance of MyClass and uses one of the properties to set the character variable to 'A'. Finally, "Is character B? False" is printed, as the call to "IsCharacterB" returns false. The static method does not return any values, so its return type is "void".

14.3 Events

Calling a method on a class causes an action to happen. In addition to causing actions within a class, we can also wait for certain events to arise from within a class. A class can expose an "event" object, which is essentially equivalent to a list of methods. Any outside instance of a class can add one of their methods to the list. When the event is triggered, each method in the list is called to notify the outside classes that the event has occurred.

```
using System;
using MyNamespace;

namespace Application
{
    class Program
    {
        static void Main(string[] args)
        {
            EventSource eventSource = new EventSource();
            EventSubscriber1 eventSubscriber1 =
                new EventSubscriber1(eventSource);
            EventSubscriber2 eventSubscriber2 =
                new EventSubscriber2(eventSource);

            Console.WriteLine("Press a key a few times");
            Console.WriteLine("or press Escape to exit.");

            while (Console.ReadKey().Key != ConsoleKey.Escape)
            {
                eventSource.Tick();
            }
        }
    }
}

namespace MyNamespace
{
    class EventSource
    {
        public delegate void MyEventHandler(int number);
        public event MyEventHandler MyEvent;

        private int countdown = 10;
        private int eventNumber = 0;

        public void Tick()
        {
            countdown = countdown - 1;
            if (countdown == 0)
            {
                countdown = 10;
                eventNumber = eventNumber + 1;

                if (MyEvent == null)
                {
                    // nobody has subscribed
                    return;
                }
```

```
                MyEvent(eventNumber);
            }
        }
    }
    class EventSubscriber1
    {
        public EventSubscriber1(EventSource eventSource)
        {
            eventSource.MyEvent += HandleEvent;
        }

        public void HandleEvent(int number)
        {
            Console.WriteLine();
            Console.WriteLine();
            Console.WriteLine("Subscriber2 got event #" + number);
        }
    }
    class EventSubscriber2
    {
        public EventSubscriber2(EventSource eventSource)
        {
            eventSource.MyEvent += OnEvent;
        }

        public void OnEvent(int number)
        {
            Console.WriteLine("Subscriber1 got event #" number + " too!");
            Console.WriteLine();
        }
    }
}
```

There is a lot of new code here, so we will address it all, beginning at the top. This code includes a "using MyNamespace;" statement, so we can omit the "MyNamespace." at the start of class names that belong to that namespace. An instance of the "EventSource" is created, followed by an instance of each of the event-subscriber classes. The event-subscriber classes are given a reference to the event source class, so they can subscribe to its event. The while loop then reads a key from the console repeatedly until it detects the escape key has been pressed. The ReadKey function returns an object with a property named "Key" that indicates which key was pressed.

Moving on to our new classes below, we start with the EventSource. Event source creates a type of event handler, then the event itself. The event handler type specifies what kind of function can handle the event. In our case, we create a function type called "MyEventHandler", which is any function that returns nothing, and accepts one integer as an argument. Following this, the event itself is created. The event consists of the "event" keyword, followed by the type of function that can handle the event, followed by the name of the event.

The tick function simply counts down each time it is called. When the countdown reaches zero the event is triggered (and the countdown restarts at ten). Care must be taken to ensure there is actually an event handler to call, so the event is compared to "null". If "null", there are no handlers, so the event should not be triggered.

Now that we have a source of events, we can now create event handlers. Two classes are created to subscribe to the event. The constructor of each class adds a member function to the list of event handlers. These functions must match the handler type specified for the event (no return value and one integer argument).

You can now press keys several times. Each press will call the "Tick" method. After ten ticks, the event is triggered and all of the functions in the handler list are called.

14.4 Exceptions

Thus far we have assumed that all operations successfully carry out what we request. Unfortunately, methods can fail and invalid operations can be attempted. C# employs a mechanism called "exception handling" to deal with notifications that something has gone awry.

An exception is an object that is created when an exceptional (exceptionally bad) situation arises. Once created, the exception is "thrown", until some code "catches" it. Once an exception is thrown, each method exits back to its caller (similar to if a "return; statement had been inserted), starting at the method that "threw" the exception, until it reaches a point within a "try" block. The exception can then be caught and handled. If this works its way back out of the first method in the stack, the default handler will be used. This handler typically displays a message to the user informing them that something terrible has happened within the application and the developer did not bother handling it.

```
using System;
using MyNamespace;

namespace Application
{
    class Program
    {
        static void Main(string[] args)
        {
            MyClass myClass = new MyClass();

            Console.WriteLine("Catching...");
            Console.ReadKey();
            myClass.One();
```

```
                    Console.WriteLine("Not catching...");
                    Console.ReadKey();
                    myClass.Two();
                }
            }
        }

        namespace MyNamespace
        {
            class MyClass
            {
                public void One()
                {
                    try
                    {
                        int x;
                        x = 5;
                        Two();
                    }
                    catch (InvalidOperationException exception)
                    {
                        Console.WriteLine("Got an invalid operation exception");
                        Console.WriteLine("Message: " + exception.Message);
                    }
                }

                public void Two()
                {
                    int y;
                    y = 7;
                    try
                    {
                        Three();
                    }
                    catch (ArithmeticException exception)
                    {
                        Console.WriteLine("Got an arithmetic exception");
                        Console.WriteLine("Message: " + exception.Message);
                    }
                }

                public void Three()
                {
                    int z;
                    throw new InvalidOperationException("Uh oh!");
                    z = 12;
                }
            }
        }
```

Pressing a key will cause the program to call the method named "One". The first call to the method "One" causes a call to the method "Two", which calls method "Three", which throws an exception. At this point, the call stack contains four methods: "Three", called by "Two", called by "One", called by "Main".

The "throw" is not contained in a "try" block, so the method exits immediately, returning to its caller, "Two". Back inside the "Two" method, the call to "Three" is in a "try" block, but the type of exception that it catches does not match the type of exception thrown. Consequently, the "Two" method exits immediately, returning back to "One". Inside "One", the call to "Two" is inside a try block that catches the type of exception that was thrown. The corresponding catch block is carried out, handling the exception.

Pressing a key again calls the method "Two", which calls the method "Three". This time, none of the methods in the chain of calls ("Three", "Two", "Main") contain a handler for the type of exception thrown. The exception exits the main method, reaching the default handler in the runtime. If you are debugging the application, your debugger will likely catch the exception. If you run your application on its own, you will likely be presented with a message that indicates your program has crashed due to an unhandled exception.

14.5 Exercises

1. Design a function that when called with an integer argument of -1 will print 'Hello' one hundred times. Do not use any for/do/do-while loops.

2. How can you detect an exception and still allow it to travel to previous calls, so other functions can also detect it?

3. If the line 'using System.Console;' is added to the top of a file, can we simply say 'WriteLine("Hello")' to write 'Hello'?

15

Application Design

15.1 Design Overview

It is quite easy to get lost in a sea of definitions when first learning a programming language. The most practical way to become a skilled programmer is to simply practice programming. In order to illustrate some basic programming, we will create a sample application over the next few chapters. The simple CPU from the first part of this book should still be somewhat familiar material, so we will create the emulator used to test programs designed for that CPU.

The CPU was designed by interconnecting transistors to form gates, then interconnecting gates to form simple machines. Finally, the simple machines were connected to create a programmable computer. The entire CPU consisted only of transistors and wires between them. Presumably, if we can simulate the behavior of a transistors connected by wires, we can simulate the entire CPU.

The CPU and main memory can both be implemented using transistors and wires, as they were in part one of this book. The BIOS, display, and external signals (clock, reset, etc.) will be implemented directly in C#. Throughout the following sections each block of code contains one class, and can be placed in its own file. It is common practice to write one class per file, with each file named after the class it contains. You can examine the resources folders for this chapter to see the files in detail as each class is implemented.

15.2 Connectors

Connectors are simply the carriers of electricity. For example, a transistor has three connectors: input, output, and source. Connectors can be connected to each other. When any one connector switches to high voltage, all connectors that are connected to it will also switch to high voltage.

```csharp
using System;
using System.Threading;
using System.Collections.Generic;

namespace Hardware
{
    class Connector
    {
        // Create voltage-change event
        public delegate void VoltageChangedHandler();
        public event VoltageChangedHandler VoltageChanged;

        // Store current voltage
        private bool isHighVoltage = false;

        // List of connections
        private List<Connector> connections = new List<Connector>();

        public void ConnectTo(Connector connector)
        {
            // Add another connector to the list of connections,
            // and add this connector to the other connector's
            // list of connections
            connections.Add(connector);
            connector.connections.Add(this);
        }

        public bool IsHighVoltage
        {
            get { return isHighVoltage; }

            set
            {
                // See if the voltage changed
                if (isHighVoltage != value)
                {
                    // Store the new voltage
                    isHighVoltage = value;

                    // Notify and event listeners
                    // that the voltage changed
                    if (VoltageChanged != null)
                    {
                        VoltageChanged();
                    }

                    // Change the voltage of all connected items
                    for (int index = 0; index < connections.Count; index++)
                    {
                        connections[index].IsHighVoltage = value;
                    }
                }
```

```
                }
              }
            }
          }
        }
      }
```

The connector consists of three main parts: the voltage changed event, the current voltage, and the list of connections. The list of connections uses a built-in class called "List". The list class maintains a list of references of a specified type ("Connector" in this case). The "ConnectTo" method connects two connectors together by adding each connector to the other's list of connections.

The IsHighVoltage property manages the isHighVoltage member variable and raises the VoltageChanged event when a change occurs. In addition to raising the event, each voltage change enters a "for" loop to access each item in the list of connected connectors. Changes in voltage are propagated to all connected connectors.

This connection class operates under the assumption that each group of interconnected connectors has only one output switching between high voltage and low voltage, and that all other connectors are inputs. When one connector switches to low voltage it sets all other connectors to low voltage without checking for any others at high voltage. If you review the design of the CPU, this assumption is perfectly valid in all cases except two: the output of the OR gate and the output of the NAND gate. Rather than slow down all connectors, we can handle these two cases specifically with a class to "mix" the two outputs.

```
namespace Hardware
{
    class VoltageMixer
    {
        private Connector input0 = new Connector();
        private Connector input1 = new Connector();
        private Connector output = new Connector();

        public VoltageMixer()
        {
            input0.VoltageChanged += MixVoltages;
            input1.VoltageChanged += MixVoltages;
        }

        public Connector Input0
        {
            get { return input0; }
        }

        public Connector Input1
        {
            get { return input1; }
        }
```

Chapter 15

```
    public Connector Output
    {
        get { return output; }
    }

    private void MixVoltages()
    {
        output.IsHighVoltage =
            input0.IsHighVoltage | input1.IsHighVoltage;
    }
  }
}
```

The voltage mixer class simply listens for voltage changes on two input connectors, and updates the output connector. The output is high voltage if either if the inputs are true. The class exposes its connectors via properties so outside classes may connect to them.

15.3 Updating

Electrical voltages are quite simple, as they propagate immediately. Unlike electrical signals, transistors do not propagate changes immediately. There is a small delay between inputs changing and the final value appearing on the output. Although the duration of the delay is not too important for our purposes, the order in which values change is important. Simply updating transistor outputs in response to input changes will yield unrealistic behavior. Consider the following schematic.

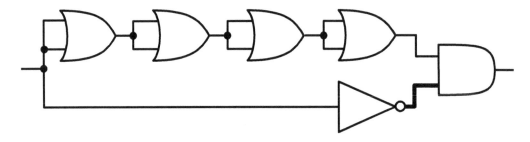

In practice, changing the input (on the left) from low voltage to high voltage should not cause the output (on the right) to change to high voltage. The NOT gate will switch to low voltage long before all of the OR gates can switch. Thus, the AND gate will never have both inputs set to high voltage.

However, consider the situation if the simulation did update each output immediately in response to an input change. When the main input on the left switches to high voltage, the OR gate inputs will switch to high voltage. Updating the output to the OR gate would update the inputs to the next OR gate, and so on. Eventually, the final OR gate would set the top input of the AND gate to high voltage, and immediately update the overall output (on the right) to high voltage, before the NOT gate is considered. This leads to incorrect behavior.

To properly simulate real-world circuits, transistor outputs need to be updated in order. In the example above, both of the inputs to the first OR gate and the input to the NOT gate would have to switch to high voltage before the output of either switched to high voltage. Similarly, the output of both the NOT gate and the first OR gate should be updated before the output of the second OR gate is updated.

A simple way to satisfy the ordering requirement is with the use of a "queue". A queue is simply a list in which items are added to the end, and remove from the front. This ensures that the first items to be added are the first items to be removed (that is, items are processed in order). Whenever the inputs on a transistor change, we will determine what the output should be, then add it to the queue of outputs to be updated. We can then examine all of the other connectors that need to change immediately, adding the consequences of their changes to the queue as well. This allows us to deal with all of the connectors that need to change immediately, while leaving the delayed transistor changes to be dealt with later. The queue will continuously be emptied, updating transistor outputs in the order they were added.

```csharp
using System.Collections.Generic;

namespace Hardware
{
    static class TransitionQueue
    {
        // Container to hold the details of a pending output change
        private class PendingTransition
        {
            public Connector Connector;
            public bool IsHighVoltage;
        }

        // Queue of pending transitions
        private static Queue<PendingTransition> pendingTransitions =
            new Queue<PendingTransition>();

        private static bool processing = false;

        public static void AddPendingTransition(Connector connector,
                                                bool isHighVoltage)
        {
            // Add new transition to the end of the queue
            PendingTransition newTransition = new PendingTransition();
            newTransition.Connector = connector;
            newTransition.IsHighVoltage = isHighVoltage;
            pendingTransitions.Enqueue(newTransition);

            if (processing)
            {
```

```
            // If we're already in a processing loop don't
            // start another
            return;
        }

        // Process pending transitions
        processing = true;
        while (pendingTransitions.Count > 0)
        {
            // Get the next pending transition
            PendingTransition transition =
                pendingTransitions.Dequeue();

            // Perform the transition
            task.Connector.IsHighVoltage =
                task.IsHighVoltage;
        }
        processing = false;
    }
  }
}
```

The first transistor to change will queue its change of output, then enter the processing loop. The processing loop will apply the change of output. Applying the output change causes the "set" accessor of the "IsHighVoltage" property to update the voltage of many connected transistors. Each of the connected transistors may add their output changes to the queue (and then exit immediately, thanks to the Boolean variable named "processing"). The "set" accessor of the "IsHighVoltage" property does not exit until all connections have been updated (which means that all connected transistors will have their output changes added to the queue). Only once all connected transistors have queued their changes will the applying of output changes continue. This ensures that all connected transistors will have their outputs updated before any further changes are processed; any output changes deeper in the circuit will be added to the end of the queue, after the current batch of output changes.

Only one update queue is needed for the application, so all of the member methods and variables have been made static. This allows all of the classes that need to use the queue to do so without keeping a reference to an instance of the queue.

15.4 Transistors

The final basic unit is the transistor. There are two types of transistors in our CPU: the normal type and the inverting type. Each type will have three connectors, and will determine its output based on the source and input.

```
namespace Hardware
{
    class NormalTransistor
    {
        private Connector source = new Connector();
        private Connector input = new Connector();
        private Connector output = new Connector();

        public NormalTransistor()
        {
            source.VoltageChanged += UpdateOutput;
            input.VoltageChanged += UpdateOutput;
            UpdateOutput();
        }

        public Connector Source
        {
            get { return source; }
        }

        public Connector Input
        {
            get { return input; }
        }

        public Connector Output
        {
            get { return output; }
        }

        private void UpdateOutput()
        {
            TransitionQueue.AddPendingTransition(output,
                source.IsHighVoltage && input.IsHighVoltage);
        }
    }
}
```

The transistor class is quite simple. Aside from creating the three connectors (and exposing them via properties), the class updates its output whenever one of the inputs changes. Recall that returning the connectors (e.g. "return output;") returns a reference to that particular instance of the Connector class. Whatever code receives a reference to the instance may then access its member methods, properties, events, etc.

The transistor class listens for voltage change events on each of its inputs. Whenever a change occurs, an output change is added to the queue of pending changes. The inverting transistor is almost identical, but produces the opposite output.

```
namespace Hardware
{
    class InvertingTransistor
    {
        private Connector source = new Connector();
        private Connector input = new Connector();
        private Connector output = new Connector();

        public InvertingTransistor()
        {
            source.VoltageChanged += UpdateOutput;
            input.VoltageChanged += UpdateOutput;
            UpdateOutput();
        }

        public Connector Source
        {
            get { return source; }
        }

        public Connector Input
        {
            get { return input; }
        }

        public Connector Output
        {
            get { return output; }
        }

        private void UpdateOutput()
        {
            TransitionQueue.AddPendingTransition(output,
                source.IsHighVoltage && !input.IsHighVoltage);
        }
    }
}
```

All of the basic components that form the CPU have now been implemented. All of the remaining parts are simply combinations of transistors, inverting transistors, and connections between them. The following classes will not actually implement the functionality of the device they represent, but will implement the ability to assemble that device using transistors and connectors.

15.5 Gates

We can now connect transistors to form the next layer of the system: gates. Each gate will be formed by instantiating and connecting transistors according to the schematics in part one of this book.

The AND gate is simply two normal transistors with the output of one connected to the source of the other. Both transistor inputs and the remaining output are exposed as external connections. Recall that although omitted in many schematics, the power source connection is required, and should also be exposed.

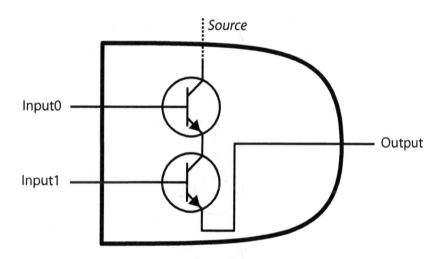

```
namespace Hardware
{
    class ANDGate
    {
        private NormalTransistor transistor0 = new NormalTransistor();
        private NormalTransistor transistor1 = new NormalTransistor();

        public ANDGate()
        {
            transistor0.Output.ConnectTo(transistor1.Source);
        }

        public Connector Source
        {
            get { return transistor0.Source; }
        }

        public Connector Input0
        {
            get { return transistor0.Input; }
        }
```

6

```
        public Connector Input1
        {
            get { return transistor1.Input; }
        }

        public Connector Output
        {
            get { return transistor1.Output; }
        }
    }
}
```

Each instance of the ANDGate class will create two instances of the NormalTransistor class and make the connection between the output and source. All remaining connectors are exposed as properties.

The OR gate is similar, but has two outputs that must be merged to create a single output. This requires the use of the voltage-mixer class that we created earlier.

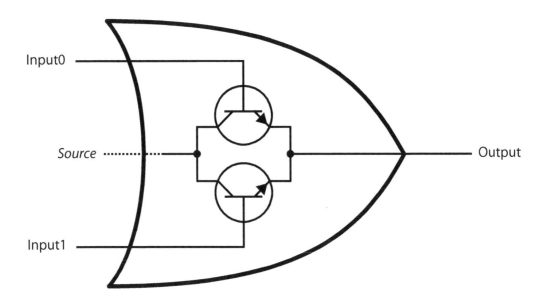

```
namespace Hardware
{
    class ORGate
    {
        private NormalTransistor transistor0 = new NormalTransistor();
        private NormalTransistor transistor1 = new NormalTransistor();
        private Connector source = new Connector();
        private VoltageMixer outputMixer = new VoltageMixer();

        public ORGate()
        {
            source.ConnectTo(transistor0.Source);
            source.ConnectTo(transistor1.Source);
            transistor0.Output.ConnectTo(outputMixer.Input0);
            transistor1.Output.ConnectTo(outputMixer.Input1);
        }

        public Connector Source
        {
            get { return source; }
        }

        public Connector Input0
        {
            get { return transistor0.Input; }
        }

        public Connector Input1
        {
            get { return transistor1.Input; }
        }

        public Connector Output
        {
            get { return outputMixer.Output; }
        }
    }
}
```

In addition to the transistors and voltage mixer, this class creates a connector object. This new connector named "source" is simply to make using the object easier. External classes that wish to use the OR gate should be able to simply connect the two inputs, the output, and the power source. The OR gate needs the power source to be connected to both transistors. Rather than require external classes to make both of these connections, we can create the single "source" connector and connect it to both transistors. This single connection can then be exposed, and will propagate any of its voltage changes to both transistors inside the OR gate.

The NOR and NAND gate schematics are identical to the schematics for the AND and OR gates, with inverting transistors in place of normal transistors. Consequently, the NOR and NAND classes are virtually identical to the AND and OR classes.

First we will create the NOR gate class, the inverting transistor version of the AND gate.

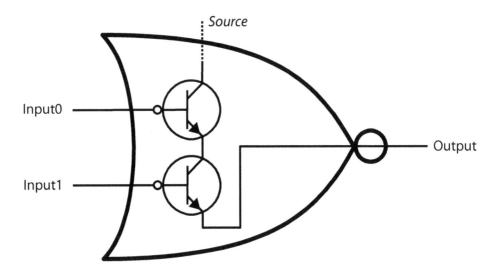

```
namespace Hardware
{
    class NORGate
    {
        private InvertingTransistor transistor0 =
            new InvertingTransistor();
        private InvertingTransistor transistor1 =
            new InvertingTransistor();

        public NORGate()
        {
            transistor0.Output.ConnectTo(transistor1.Source);
        }

        public Connector Source
        {
            get { return transistor0.Source; }
        }

        public Connector Input0
        {
            get { return transistor0.Input; }
        }

        public Connector Input1
        {
            get { return transistor1.Input; }
        }

        public Connector Output
        {
            get { return transistor1.Output; }
        }
    }
}
```

Next is the NAND gate class, the inverting transistor version of the OR gate. It also contains an extra source connector that is internally connected to both transistors.

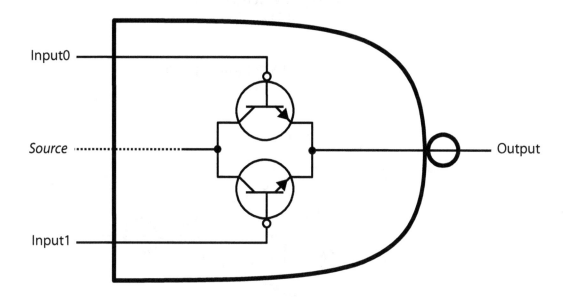

```
namespace Hardware
{
    class NANDGate
    {
        private InvertingTransistor transistor0 =
            new InvertingTransistor();
        private InvertingTransistor transistor1 =
            new InvertingTransistor();
        private Connector source = new Connector();
        private VoltageMixer outputMixer = new VoltageMixer();

        public NANDGate()
        {
            source.ConnectTo(transistor0.Source);
            source.ConnectTo(transistor1.Source);
            transistor0.Output.ConnectTo(outputMixer.Input0);
            transistor1.Output.ConnectTo(outputMixer.Input1);
        }

        public Connector Source
        {
            get { return source; }
        }

        public Connector Input0
        {
            get { return transistor0.Input; }
        }

        public Connector Input1
        {
            get { return transistor1.Input; }
        }
```

```
        public Connector Output
        {
            get { return outputMixer.Output; }
        }
    }
}
```

The final two gates are the NOT and XOR gate. The NOT gate requires only a single inverting transistor.

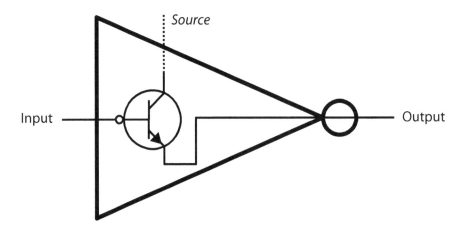

```
namespace Hardware
{
    class NOTGate
    {
        private InvertingTransistor transistor =
            new InvertingTransistor();

        public Connector Source
        {
            get { return transistor.Source; }
        }

        public Connector Input
        {
            get { return transistor.Input; }
        }

        public Connector Output
        {
            get { return transistor.Output; }
        }
    }
}
```

The XOR gate was implemented using other gates, rather than directly using transistors. The gates we have just implemented expose the same connector objects that were used in the transistors. Hence, gates can be connected in the same manner as the transistors. Referring back to the XOR gate schematics shows that it requires a NAND gate, an OR gate, and an AND gate.

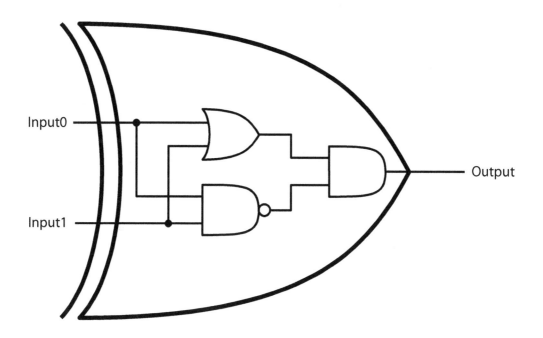

```
namespace Hardware
{
    class XORGate
    {
        private NANDGate nand = new NANDGate();
        private ORGate or = new ORGate();
        private ANDGate and = new ANDGate();

        private Connector input0 = new Connector();
        private Connector input1 = new Connector();
        private Connector source = new Connector();

        public XORGate()
        {
            input0.ConnectTo(or.Input0);
            input0.ConnectTo(nand.Input0);

            input1.ConnectTo(or.Input1);
            input1.ConnectTo(nand.Input1);

            or.Output.ConnectTo(and.Input0);
            nand.Output.ConnectTo(and.Input1);

            source.ConnectTo(or.Source);
            source.ConnectTo(nand.Source);
            source.ConnectTo(and.Source);
        }
```

```
    public Connector Source
    {
        get { return source; }
    }

    public Connector Input0
    {
        get { return input0; }
    }

    public Connector Input1
    {
        get { return input1; }
    }

    public Connector Output
    {
        get { return and.Output; }
    }
}
}
```

There are a few more connections in the XOR gate than in previous gates, but the concept is the same. Each gate is added and connected according to the schematic for the XOR gate (as well as power to each gate, which was omitted in the schematics for the sake of clarity). Extra connectors are made for the source and for each input, as all of these signals must be sent to multiple internal components. These connectors allow us to expose a single connection that will be sent to multiple places within the XOR gate class.

Whenever a XOR gate is instantiated, a NAND gate, OR gate, and AND gate are created. Each of these three gates creates two transistors. Therefore, each instance of the XOR gate creates six transistors, as well as setting up the appropriate connections to obtain the exclusive-or behavior. After all the connections have been established, only the two XOR inputs, one XOR output, and a power connection remain. All three of these are exposed as the connectors to the XOR gate.

15.6 Exercises

1. The List<*type*> class is very similar to the Queue<*type*> class. How could the update queue class be implemented using a List instead of a Queue (and retain the original functionality)?

2. When a connector's voltage changes, it propagates its value to all connected connectors. Why does the program not get stuck in an infinite loop when this happens? e.g. connector1 updates connector2, which updates connector1, which updates connector2, etc.

3. Is this design the fastest way to carry out the instructions for the sample CPU?

16

Components

16.1 Multiplexer

With the simple logic gates fully implemented, we can now begin assembling some simple machines. One of the simpler gate-based devices is a two-input single-bit multiplexer.

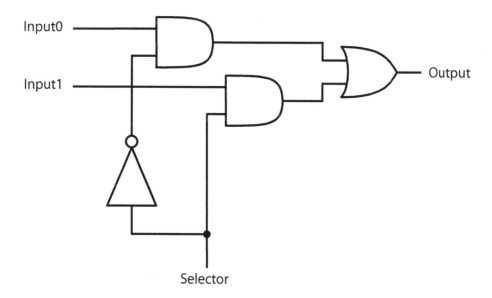

```
namespace Hardware
{
    class SingleBitMultiplexer
    {
        private ANDGate and0 = new ANDGate();
        private ANDGate and1 = new ANDGate();
        private ORGate or = new ORGate();
        private NOTGate not = new NOTGate();

        private Connector source = new Connector();
        private Connector selector = new Connector();

        public SingleBitMultiplexer()
        {
            and0.Output.ConnectTo(or.Input0);
            and1.Output.ConnectTo(or.Input1);

            selector.ConnectTo(not.Input);
            not.Output.ConnectTo(and0.Input1);
            selector.ConnectTo(and1.Input1);

            source.ConnectTo(and0.Source);
            source.ConnectTo(and1.Source);
            source.ConnectTo(or.Source);
            source.ConnectTo(not.Source);
        }

        public Connector Input0
        {
            get { return and0.Input0; }
        }

        public Connector Input1
        {
            get { return and1.Input0; }
        }

        public Connector Output
        {
            get { return or.Output; }
        }

        public Connector Selector
        {
            get { return selector; }
        }

        public Connector Source
        {
            get { return source; }
        }
    }
}
```

Aside from merging the multiple source and selector connectors into a single convenient connector, this code simply creates and connects several basic gates according to the schematic.

Because the code simulates the actual hardware, we can use the same shortcuts we used when implementing the hardware. In this case, a multi-bit two-input multiplexer can be created by connecting multiple single-bit two-input multiplexers.

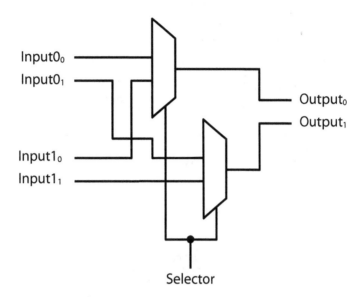

Although the schematic shows a two-bit implementation, the same pattern can be extended to any number of bits. Rather than decide on a fixed number of bits, we can create a more versatile multi-bit multiplexer. An argument to the constructor is a convenient place to specify the number of bits. With a size specified, we can simply create an array of single-bit multiplexers. One multiplexer is required for each bit of the number. The top input of each multiplexer is one bit of the first multi-bit input, while the bottom input of each multiplexer is one bit of the second multi-bit input. Each output from the multiplexers is a single bit of the multi-bit output.

```
namespace Hardware
{
    class MultiBitMultiplexer
    {
        SingleBitMultiplexer[] multiplexers;

        Connector selector = new Connector();
        Connector source = new Connector();

        public MultiBitMultiplexer(int size)
        {
            multiplexers = new SingleBitMultiplexer[size];
            for (int index = 0; index < size; index++)
            {
                multiplexers[index] = new SingleBitMultiplexer();
                source.ConnectTo(multiplexers[index].Source);
                selector.ConnectTo(multiplexers[index].Selector);
```

```
        }
    }

    public Connector Source
    {
        get { return source; }
    }

    public Connector[] Input0
    {
        get
        {
            Connector[] connectors =
                new Connector[multiplexers.Length];
            for (int index = 0; index < multiplexers.Length; index++)
            {
                connectors[index] = multiplexers[index].Input0;
            }
            return connectors;
        }
    }

    public Connector[] Input1
    {
        get
        {
            Connector[] connectors =
                new Connector[multiplexers.Length];
            for (int index = 0; index < multiplexers.Length; index++)
            {
                connectors[index] = multiplexers[index].Input1;
            }
            return connectors;
        }
    }

    public Connector[] Output
    {
        get
        {
            Connector[] connectors =
                new Connector[multiplexers.Length];
            for (int index = 0; index < multiplexers.Length; index++)
            {
                connectors[index] = multiplexers[index].Output;
            }
            return connectors;
        }
    }

    public Connector Selector
    {
        get { return selector; }
    }
  }
}
```

The constructor creates an array of SingleBitMultiplexer references using the size argument. A "for" loop then creates each SingleBitMultiplexer and stores its reference in the array. The loop also connects the power source to each single-bit multiplexer.

All of the selector connectors on the single-bit multiplexers are connected to a single exposed selector connector for convenient use by external classes. The same is done for the power-source connectors. Unlike these connections, the multiple input connectors and the output connectors cannot be merged. These connectors can (and must) carry different values to properly transmit a multi-bit number. Consequently, we must return a group of connectors for the inputs and output. Each of the input and output properties create an array of connector references, and then populate the array with the input connectors or output connectors.

We can now move on to a more complicated device: the generic multi-input multi-bit multiplexer. Understanding the implementation of this class is heavily dependent on understanding the multiplexer design from the first part of the book. The most complicated part of this class is the constructor. This first thing you may notice is the strange array. Accepting a larger number of inputs requires multiple columns of two-input multiplexers. The pattern that these columns follow is the key to the multiplexer's functionality.

- Each bit of the selector is fed into an entire column of multiplexers
- Each column (and each bit of the selector) halves the number of inputs
- Each column has half as many multiplexers as the previous (to the left)
- Numbering from the top of a column, starting with zero, the output from multiplexer number "x" feeds into multiplexer number "x / 2" in the next column (to the right). For example, the output from multiplexers number 4 and 5 feed into multiplexer number 2 in the next column
- The rightmost column has a single multiplexer that narrows the inputs down to the single output
- The leftmost column accepts all of the inputs, requiring half as many multiplexers as there are inputs (as each multiplexer accepts two of the inputs).

The sixteen-to-one multiplexer illustrates this pattern.

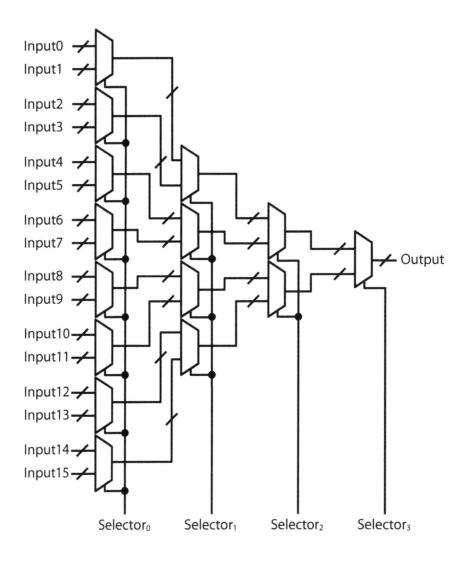

```
namespace Hardware
{
    class Multiplexer
    {
        MultiBitMultiplexer[][] multiplexerColumns;

        Connector source = new Connector();
        Connector[] selector;

        private int selectorSize;
        private int dataSize;

        public Multiplexer(int dataSize, int selectorSize)
        {
            // Copy arguments to class member variables
```

```
        this.dataSize = dataSize;
        this.selectorSize = selectorSize;

        // Create an array of multi-bit multiplexer arrays
        multiplexerColumns = new MultiBitMultiplexer[selectorSize][];

        // Create an array of selectors (one for each
        // bit of the slector)
        selector = new Connector[selectorSize];

        // Keep track of how many multiplexers should go
        // in the current column starting with the last column
        int columnMultiplexerCount = 1;
        for (int columnIndex = selectorSize - 1; columnIndex >= 0;
            columnIndex--)
        {
            // Create a column and store it in the array
            MultiBitMultiplexer[] multiplexerColumn =
                new MultiBitMultiplexer[columnMultiplexerCount];
            multiplexerColumns[columnIndex] = multiplexerColumn;

            // Create the selector bit for this column
            Connector selectorBit = new Connector();
            selector[columnIndex] = selectorBit;

            // Create each of the multiplexers in the column
            for (int multiplexerIndex = 0;
                multiplexerIndex < columnMultiplexerCount;
                multiplexerIndex++)
            {
                // Create the multiplexer and store it in
                // the current column
                MultiBitMultiplexer multiplexer =
                    new MultiBitMultiplexer(dataSize);
                multiplexerColumn[multiplexerIndex] = multiplexer;

                // Connect the multiplexer to the power and selector
                source.ConnectTo(multiplexer.Source);
                selectorBit.ConnectTo(multiplexer.Selector);

                // See if this is any column except the last
                if (columnIndex < selectorSize - 1)
                {
                    // Get the next column
                    MultiBitMultiplexer[] nextMultiplexerColumn =
                        multiplexerColumns[columnIndex + 1];

                    Connector[] outputDestination;

                    // See if this is an even numbered multiplexer
                    // or odd numbered multiplexer
                    if((multiplexerIndex & 1) == 0)
                    {
                        // Even numbered multiplexers connect to the
                        // top input
                        outputDestination =
                            nextMultiplexerColumn[multiplexerIndex
                                / 2].Input0
                    }
                    else
                    {
                        // Odd numbered multiplexers connect to the
                        // bottom input
```

```
                    outputDestination =
                        nextMultiplexerColumn[multiplexerIndex
                            / 2].Input1;
                }

                // Now that we have matched this output to the input
                // in the next column, connect each bit of the
                // output to each bit of the input
                for (int dataIndex = 0; dataIndex < dataSize;
                    dataIndex++)
                {
                    multiplexer.Output[dataIndex].ConnectTo(
                        outputDestination[dataIndex]);
                }
            }
        }

        // The previous column needs twice as many multiplexers as
        // this column
        columnMultiplexerCount *= 2;
    }
}

public Connector[][] Inputs
{
    get
    {
        // All inputs come from the first column
        MultiBitMultiplexer[] firstColumn = multiplexerColumns[0];

        // Each multiplexer has two inputs, so we need room for
        // twice as many connectors as there are multiplexers
        Connector[][] connectors = new Connector[firstColumn.Length
            * 2][];
        for (int index = 0; index < firstColumn.Length; index++)
        {
            connectors[index * 2] = firstColumn[index].Input0;
            connectors[index * 2 + 1] = firstColumn[index].Input1;
        }
        return connectors;
    }
}

public Connector[] Output
{
    // The outputs are simply all of the outputs from the
    // single multiplexer in the last column
    get { return multiplexerColumns[selectorSize - 1][0].Output; }
}

public Connector[] Selector
{
    get { return selector; }
}

public Connector Source
{
    get { return source; }
}
    }
}
```

The code uses a "for" loop to create one column for each bit of the selector, as well as creating the connector for each bit the selector. A second, contained "for" loop creates each multiplexer in the column and connects it to the current selector connector. For each column except the rightmost, the output of each multiplexer is connected to the input of a multiplexer in the next column.

The output property returns the array of connectors that make up the single multi-bit output value. As there are many multi-bit inputs, the inputs property returns an array of arrays of connectors.

One of the best ways to understand complicated code is to trace through it manually. The large multiplexer in the previous image would be created with a selector size of four, and any data size. You can try manually drawing out the connections that the code will create when called with a selectorSize argument of four. You should end up with the sixteen-input multiplexer. With a bit of careful examination, you can see that we now have a class capable of assembling a generic multiplexer, with a specified number of inputs (via the selector size) and a specified number of bits.

16.2 Demultiplexer

Unfortunately, there is one more tricky class to be made. The demultiplexer has a structure that is similar to the multiplexer. The single-bit, two-input demultiplexer is still only a few gates.

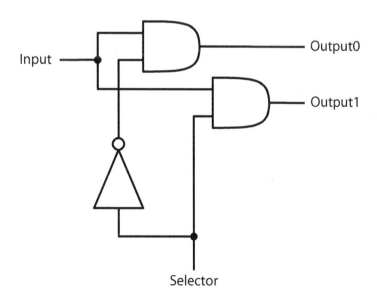

```
namespace Hardware
{
    class SingleBitDemultiplexer
    {
        private ANDGate and0 = new ANDGate();
        private ANDGate and1 = new ANDGate();
        private NOTGate not = new NOTGate();

        private Connector input = new Connector();
        private Connector source = new Connector();
        private Connector selector = new Connector();

        public SingleBitDemultiplexer()
        {
            input.ConnectTo(and0.Input0);
            input.ConnectTo(and1.Input0);

            selector.ConnectTo(not.Input);
            not.Output.ConnectTo(and0.Input1);
            selector.ConnectTo(and1.Input1);

            source.ConnectTo(and0.Source);
            source.ConnectTo(and1.Source);
            source.ConnectTo(not.Source);
        }

        public Connector Input
        {
            get { return input; }
        }

        public Connector Output0
        {
            get { return and0.Output; }
        }

        public Connector Output1
        {
            get { return and1.Output; }
        }

        public Connector Selector
        {
            get { return selector; }
        }

        public Connector Source
        {
            get { return source; }
        }
    }
}
```

Once again, using multiple single-bit devices can create a multi-bit device. Each demultiplexer chooses which of the two outputs will receive a particular bit from the input.

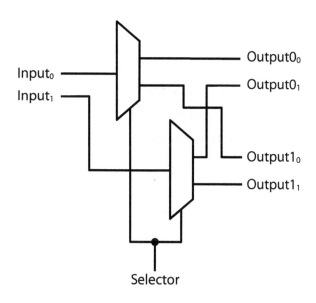

```
namespace Hardware
{
    class MultiBitDemultiplexer
    {
        SingleBitDemultiplexer[] demultiplexers;

        Connector selector = new Connector();
        Connector source = new Connector();

        public MultiBitDemultiplexer(int size)
        {
            demultiplexers = new SingleBitDemultiplexer[size];
            for (int index = 0; index < size; index++)
            {
                demultiplexers[index] = new SingleBitDemultiplexer();
                source.ConnectTo(demultiplexers[index].Source);
                selector.ConnectTo(demultiplexers[index].Selector);
            }
        }

        public Connector Source
        {
            get { return source; }
        }

        public Connector[] Input
        {
            get
            {
                Connector[] connectors =
                    new Connector[demultiplexers.Length];
```

```
                    for (int index = 0; index < demultiplexers.Length; index++)
                    {
                        connectors[index] = demultiplexers[index].Input;
                    }
                    return connectors;
                }
            }

            public Connector[] Output0
            {
                get
                {
                    Connector[] connectors =
                        new Connector[demultiplexers.Length];
                    for (int index = 0; index < demultiplexers.Length; index++)
                    {
                        connectors[index] = demultiplexers[index].Output0;
                    }
                    return connectors;
                }
            }

            public Connector[] Output1
            {
                get
                {
                    Connector[] connectors =
                        new Connector[demultiplexers.Length];
                    for (int index = 0; index < demultiplexers.Length; index++)
                    {
                        connectors[index] = demultiplexers[index].Output1;
                    }
                    return connectors;
                }
            }

            public Connector Selector
            {
                get { return selector; }
            }

        }
    }
```

Finally, the generic multi-bit, multi-output demultiplexer. This device follows the same type of pattern as the generic multiplexer.

Columns of devices are created, with the outputs of one column connected to the inputs of the next column. Each bit of the selector controls one column.

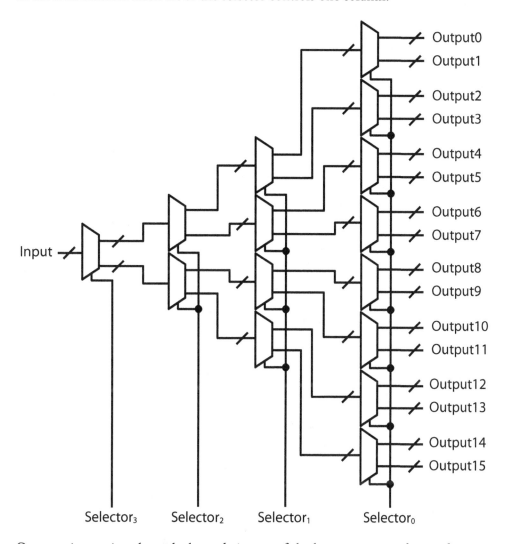

Once again, tracing through the code is one of the best ways to understand its operation. The demultiplexer shown above is created with any data size, and a selector size of four. The implementation is very similar to the implementation of the generic multiplexer.

```
namespace Hardware
{
    class Demultiplexer
    {
        MultiBitDemultiplexer[][] columns;

        Connector source = new Connector();
        Connector[] selector;

        private int selectorSize;
        private int dataSize;

        public Demultiplexer(int dataSize, int selectorSize)
        {
            // Copy arguments into the class member variables
            this.dataSize = dataSize;
            this.selectorSize = selectorSize;

            // Create the array of columns of two-output single-bit
            // demultiplexers
            columns = new MultiBitDemultiplexer[selectorSize][];

            // Create the selector connector array
            selector = new Connector[selectorSize];

            // Create each column, starting at the left
            int columnDemultiplexerCount = 1;
            for (int columnIndex = 0; columnIndex < selectorSize;
                columnIndex++)
            {
                // Create the array of demultiplexer references for
                // the column
                MultiBitDemultiplexer[] demultiplexerColumn =
                    new MultiBitDemultiplexer[columnDemultiplexerCount];
                columns[columnIndex] = demultiplexerColumn;

                // Create the selector connector for the column
                Connector selectorBit = new Connector();
                selector[selectorSize - columnIndex - 1] = selectorBit;

                // Create each demultiplexer in the column
                for (int demultiplexerIndex = 0;
                    demultiplexerIndex < columnDemultiplexerCount;
                    demultiplexerIndex++)
                {
                    // Create the demultiplexer
                    MultiBitDemultiplexer demultiplexer =
                        new MultiBitDemultiplexer(dataSize);
                    demultiplexerColumn[demultiplexerIndex] =
                        demultiplexer;

                    // Connect the source and selector for this column
                    source.ConnectTo(demultiplexer.Source);
                    selectorBit.ConnectTo(demultiplexer.Selector);

                    if (columnIndex > 0)
                    {
                        // All columns except the first have their input
                        // connected to the previous column
                        MultiBitDemultiplexer[] previousDemultiplexerColumn
                            = columns[columnIndex - 1];

                        // Figure out where the input should come from
```

```
                    Connector[] inputOrigin;
                    if ((demultiplexerIndex & 1) == 0)
                    {
                        // Even numbered inputs come from the top output
                        inputOrigin = previousDemultiplexerColumn[
                            demultiplexerIndex / 2].Output0;
                    }
                    else
                    {
                        // Odd numbered inputs come from the
                        // bottom output
                        inputOrigin = previousDemultiplexerColumn[
                            demultiplexerIndex / 2].Output1;
                    }

                    // Connect each bit of the input to each bit of the
                    // previous output
                    for (int dataIndex = 0;dataIndex < dataSize;
                        dataIndex++)
                    {
                        demultiplexer.Input[dataIndex].ConnectTo(
                            inputOrigin[dataIndex]);
                    }
                }
            }

            // The next column has twice as many demultiplexers
            // as the current column
            columnDemultiplexerCount *= 2;
        }
    }

    public Connector[] Input
    {
        // The first (and only) demultiplexer accepts
        // all of the input bits
        get { return columns[0][0].Input; }
    }

    public Connector[][] Outputs
    {
        get
        {
            // All of the multi-bit outputs come from the
            // rightmost column
            MultiBitDemultiplexer[] lastColumn =
                columns[selectorSize - 1];

            // Create an array of connector arrays
            Connector[][] connectors =
                new Connector[lastColumn.Length * 2][];

            // Store both of the multi-bit outputs from
            // each demultiplexer in the rightmost column
            for (int index = 0; index < lastColumn.Length; index++)
            {
                connectors[index * 2] = lastColumn[index].Output0;
                connectors[index * 2 + 1] = lastColumn[index].Output1;
            }
            return connectors;
        }
    }
}
```

```
public Connector[] Selector
{
    get { return selector; }
}

public Connector Source
{
    get { return source; }
}
    }
}
```

16.3 Memory

With the multiplexer and demultiplexer out of the way, we can move on to implementing memory devices. The simplest memory device is the one-bit memory unit. Following the schematic, we can see that a single bit of memory requires four NAND gates, with several connections. NAND gates zero through three represent the gates in the schematic found in the top left, bottom left, top right, and bottom right, respectively.

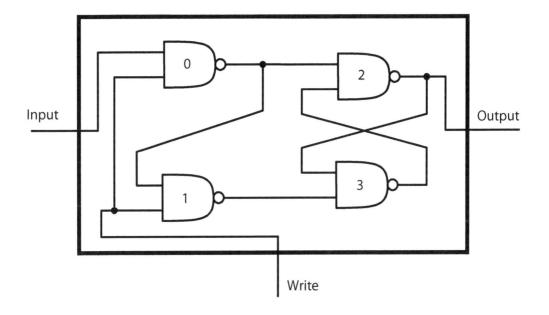

```
namespace Hardware
{
    class MemoryUnit
    {
        private Connector write = new Connector();
        private Connector source = new Connector();

        private NANDGate nand0 = new NANDGate();
        private NANDGate nand1 = new NANDGate();
        private NANDGate nand2 = new NANDGate();
        private NANDGate nand3 = new NANDGate();

        public MemoryUnit()
        {
            write.ConnectTo(nand0.Input1);
            write.ConnectTo(nand1.Input1);

            nand0.Output.ConnectTo(nand2.Input0);
            nand0.Output.ConnectTo(nand1.Input0);
            nand1.Output.ConnectTo(nand3.Input1);
            nand2.Output.ConnectTo(nand3.Input0);
            nand3.Output.ConnectTo(nand2.Input1);

            source.ConnectTo(nand0.Source);
            source.ConnectTo(nand1.Source);
            source.ConnectTo(nand2.Source);
            source.ConnectTo(nand3.Source);
        }

        public Connector Source
        {
            get { return source; }
        }

        public Connector Input
        {
            get { return nand0.Input0; }
        }

        public Connector Write
        {
            get { return write; }
        }

        public Connector Output
        {
            get { return nand2.Output; }
        }
    }
}
```

Of course, the next logical step is to create the multi-bit memory unit. The multi-bit register is created by packaging one single-bit memory unit per bit of storage. All of the memory units share the same "write" connection and "source" connection.

The following schematic shows a size of four.

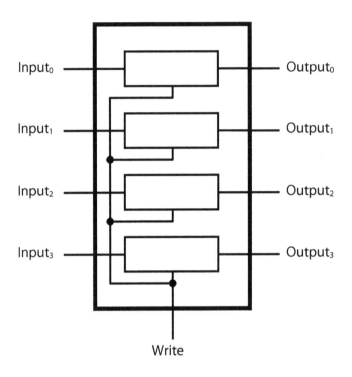

Write

```
namespace Hardware
{
    class Register
    {
        MemoryUnit[] memoryUnits;

        Connector write = new Connector();
        Connector source = new Connector();

        public Register(int size)
        {
            memoryUnits = new MemoryUnit[size];

            for (int index = 0; index < size; index++)
            {
                MemoryUnit memoryUnit = new MemoryUnit();
                source.ConnectTo(memoryUnit.Source);
                write.ConnectTo(memoryUnit.Write);
                memoryUnits[index] = memoryUnit;
            }
        }

        public Connector Source
        {
            get { return source; }
        }

        public Connector[] Input
        {
```

```
        get
        {
            Connector[] inputs = new Connector[memoryUnits.Length];
            for (int index = 0; index < memoryUnits.Length; index++)
            {
                inputs[index] = memoryUnits[index].Input;
            }
            return inputs;
        }
    }

    public Connector[] Output
    {
        get
        {
            Connector[] outputs = new Connector[memoryUnits.Length];
            for (int index = 0; index < memoryUnits.Length; index++)
            {
                outputs[index] = memoryUnits[index].Output;
            }
            return outputs;
        }
    }

    public Connector Write
    {
        get { return write; }
    }
}
}
```

The constructor creates an array of MemoryUnit references using the size argument. A "for" loop then creates each MemoryUnit and stores its reference in the array. The loop also connects the power source and the write-enable connector to each memory unit. The input property and output property each return an array of references to connectors that carry each bit of the value.

Now that we have the device to store a multi-bit number we can create a device that can store many multi-bit numbers: the memory bank. The memory bank consists of one register for each addressable location. A demultiplexer directs the write-enable signal to the appropriate register, while two multiplexers select which locations should be forwarded to the two value-output connectors.

The following memory bank can store four values (of an unspecified value size).

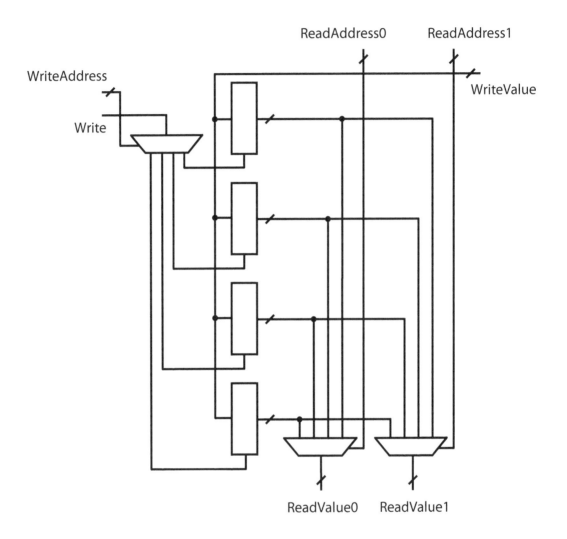

This memory bank requires both a data size and address size to determine how many registers are required, and how many bits each register should store.

```
namespace Hardware
{
    class MemoryBank
    {
        Register[] registers;
        Demultiplexer demultiplexer;
        Multiplexer multiplexer0;
        Multiplexer multiplexer1;

        Connector source = new Connector();
        Connector[] writeValue;

        public MemoryBank(int dataSize, int addressSize)
        {
            // A quick trick to calculate a power of two
            // 1 shifted left x times is equal to two to
            // the power of x.
            int registerCount = 1 << addressSize;

            // Create an array of references to hold the registers
            registers = new Register[registerCount];

            // Create a demultiplexer with one output per register.
            // The demultiplexer only carries a 1-bit enable signal
            demultiplexer = new Demultiplexer(1, addressSize);

            // Create multiplexers with one input per register.
            // The multiplexers carry the values from the register,
            // so they must be the same size as the register data.
            multiplexer0 = new Multiplexer(dataSize, addressSize);
            multiplexer1 = new Multiplexer(dataSize, addressSize);

            source.ConnectTo(demultiplexer.Source);
            source.ConnectTo(multiplexer0.Source);
            source.ConnectTo(multiplexer1.Source);

            // Create each write-value input connector and
            // store them in an array
            writeValue = new Connector[dataSize];
            for (int dataIndex = 0; dataIndex < dataSize; dataIndex++)
            {
                writeValue[dataIndex] = new Connector();
            }

            // Connect each register
            for (int registerIndex = 0; registerIndex < registerCount;
                registerIndex++)
            {
                // Create a register
                Register register = new Register(dataSize);
                registers[registerIndex] = register;
                register.Source.ConnectTo(source);

                // Get the appropriate connectors from the multieplexers
                // and demultiplexer
                Connector demultiplexerOutput =
                    demultiplexer.Outputs[registerIndex][0];
                Connector[] multiplexer0Input =
                    multiplexer0.Inputs[registerIndex];
                Connector[] multiplexer1Input =
                    multiplexer1.Inputs[registerIndex];

                // Connect the single-bit enable connector to the
```

```
                // demultiplexer output
                register.Write.ConnectTo(demultiplexerOutput);

                for (int dataIndex = 0; dataIndex < dataSize; dataIndex++)
                {
                    // Connect each bit of the register's input to the
                    // write-value
                    register.Input[dataIndex].ConnectTo(
                        writeValue[dataIndex]);

                    // Connect each bit of the register's output to the
                    // output multiplexers
                    register.Output[dataIndex].ConnectTo(
                        multiplexer0Input[dataIndex]);
                    register.Output[dataIndex].ConnectTo(
                        multiplexer1Input[dataIndex]);
                }
            }
        }

        public Connector Source
        {
            get { return source; }
        }

        public Connector Write
        {
            get { return demultiplexer.Input[0]; }
        }

        public Connector[] WriteAddress
        {
            get { return demultiplexer.Selector; }
        }

        public Connector[] WriteValue
        {
            get { return writeValue; }
        }

        public Connector[] ReadAddress0
        {
            get { return multiplexer0.Selector; }
        }

        public Connector[] ReadAddress1
        {
            get { return multiplexer1.Selector; }
        }

        public Connector[] ReadValue0
        {
            get { return multiplexer0.Output; }
        }

        public Connector[] ReadValue1
        {
            get { return multiplexer1.Output; }
        }
    }
}
```

16.4 Adder

Another important component of the CPU design is the adder. The basic unit of adding is the "full adder", which can add two one-bit numbers, taking into account a carry-in value and a carry-out value.

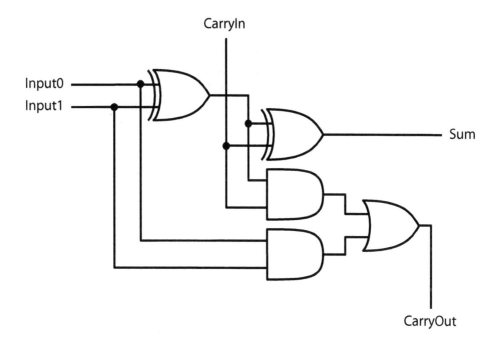

This device is simple enough. A few gates with some connections and the full adder is complete.

```
namespace Hardware
{
    class FullAdder
    {
        private XORGate xor0 = new XORGate();
        private XORGate xor1 = new XORGate();
        private ANDGate and0 = new ANDGate();
        private ANDGate and1 = new ANDGate();
        private ORGate or = new ORGate();

        private Connector source = new Connector();
        private Connector input0 = new Connector();
        private Connector input1 = new Connector();
        private Connector carryIn = new Connector();

        public FullAdder()
        {
            input0.ConnectTo(xor0.Input0);
            input0.ConnectTo(and1.Input0);
```

```
            input1.ConnectTo(xor0.Input1);
            input1.ConnectTo(and1.Input1);

            xor0.Output.ConnectTo(xor1.Input0);
            xor0.Output.ConnectTo(and0.Input0);

            carryIn.ConnectTo(xor1.Input1);
            carryIn.ConnectTo(and0.Input1);

            and0.Output.ConnectTo(or.Input0);
            and1.Output.ConnectTo(or.Input1);

            source.ConnectTo(xor0.Source);
            source.ConnectTo(xor1.Source);
            source.ConnectTo(and0.Source);
            source.ConnectTo(and1.Source);
            source.ConnectTo(or.Source);
        }

        public Connector Source
        {
            get { return source; }
        }

        public Connector Input0
        {
            get { return input0; }
        }

        public Connector Input1
        {
            get { return input1; }
        }

        public Connector Output
        {
            get { return xor1.Output; }
        }

        public Connector CarryIn
        {
            get { return carryIn; }
        }

        public Connector CarryOut
        {
            get { return or.Output; }
        }
    }
}
```

Following the trend so far, the next step is to expand this device to accommodate multi-bit values. Each adder can add one bit of the sum, so we simply need to create one adder for each bit of the desired size.

The following schematic shows a four-bit adder.

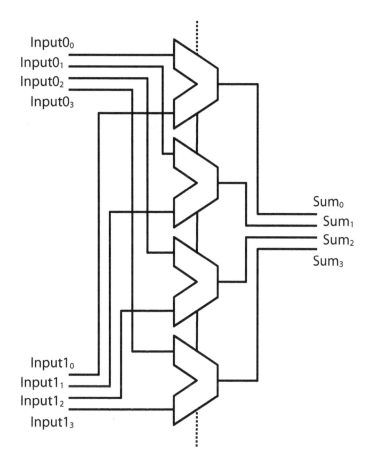

The multi-bit adder sends one bit from each of the input values to each full adder. The combined outputs from the full adders form the multi-bit sum. Each carry-out (except the last) is connected to the carry-in of the next adder.

```
namespace Hardware
{
    class Adder
    {
        private FullAdder[] adders;

        private Connector source = new Connector();

        public Adder(int size)
        {
            adders = new FullAdder[size];
            for (int index = 0; index < size; index++)
            {
                adders[index] = new FullAdder();
                source.ConnectTo(adders[index].Source);
                if (index > 0)
                {
                    adders[index - 1].CarryOut.ConnectTo(
                        adders[index].CarryIn);
                }
            }
        }

        public Connector Source
        {
            get { return source; }
        }

        public Connector[] Input0
        {
            get
            {
                Connector[] connectors = new Connector[adders.Length];
                for (int index = 0; index < adders.Length; index++)
                {
                    connectors[index] = adders[index].Input0;
                }
                return connectors;
            }
        }

        public Connector[] Input1
        {
            get
            {
                Connector[] connectors = new Connector[adders.Length];
                for (int index = 0; index < adders.Length; index++)
                {
                    connectors[index] = adders[index].Input1;
                }
                return connectors;
            }
        }

        public Connector[] Output
        {
            get
            {
                Connector[] connectors = new Connector[adders.Length];
                for (int index = 0; index < adders.Length; index++)
                {
                    connectors[index] = adders[index].Output;
                }
```

```
                    return connectors;
                }
            }
        }
    }
}
```

16.5 Delay

The final component – the delay – is the simplest. Relying on a minimum amount of time required for a transistor to transition, the delay component is simply a chain of transistors (or anything made of transistors). We will use a chain of NOT gates. For the delayed output to be the same value as the input, there must be an even number of NOT gates.

```
using System;
using System.Collections.Generic;
namespace Hardware
{
    class Delay
    {
        private Connector source = new Connector();
        private NOTGate[] gates;

        public Delay(int size)
        {
            if (size <= 0 || size % 2 != 0)
            {
                throw new ArgumentException("Invalid delay size!");
            }

            gates = new NOTGate[size];

            for (int index = 0; index < size; index++)
            {
                NOTGate not = new NOTGate();
                source.ConnectTo(not.Source);
                if (index > 0)
                {
                    gates[index - 1].Output.ConnectTo(not.Input);
                }
                gates[index] = not;
            }
        }

        public Connector Source
        {
            get { return source; }
        }

        public Connector Input
        {
            get { return gates[0].Input; }
        }

        public Connector Output
```

```
        {
            get { return gates[gates.Length - 1].Output; }
        }
    }
}
```

After ensuring the number of NOT gates is a multiple of two, the chain of gates is created and connected. This delay unit is the final component required to construct the CPU.

17

System Integration

17.1 Controller

The CPU can be logically divided into two parts: the "main" part, and the "controller". The controller is a CPU-specific device that controls the connections between the various functional parts of the CPU.

All of the previous components that have been implemented can be reused to model a variety of electronic devices. The controller is kept separate from these classes as it is specific to this CPU. As projects grow in size, it is very important to have a clean, organized hierarchy of files. You can examine the resources folder for this chapter – 17 – to see a sample of file organization. Many programming environments, including Visual C#, allow you to manage files and folders directly within the application you use to compose a program. The name and location a file does not have any impact on the functionality described within it. Once you are satisfied with the organization of the project, we can continue on to implementing the controller.

The controller for our CPU is the following combination of gates.

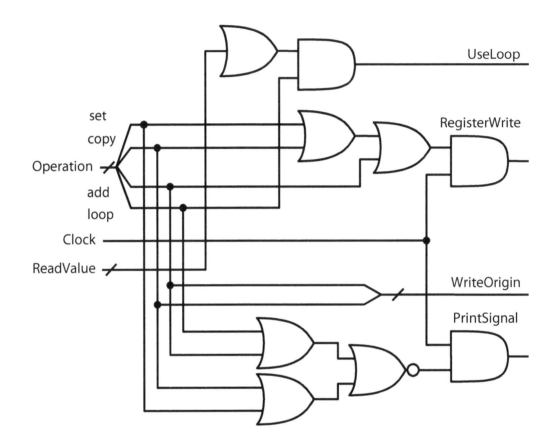

All of the connections on the left side are inputs. The logical combinations of the inputs generate the outputs that control the rest of the CPU.

Recall that the topmost OR gate represents a large sixteen-input OR gate, as follows.

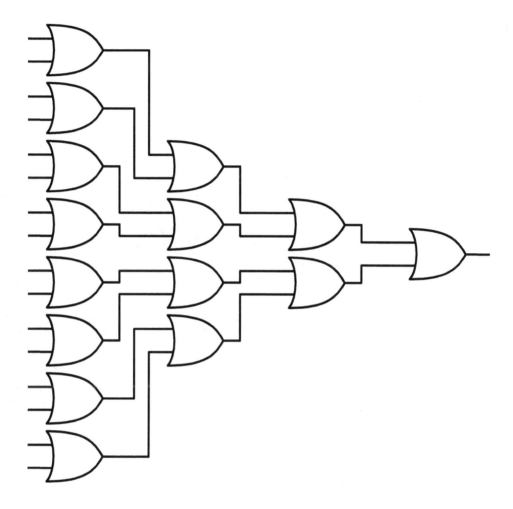

All of the other gates are the basic two-input variety, making this a rather simple device.

```
namespace Hardware
{
    class Controller
    {
        private Connector source = new Connector();
        private ORGate[] readValueORs = new ORGate[15];
        private ORGate operationOR0 = new ORGate();
        private ORGate operationOR1 = new ORGate();
        private NORGate operationNOR = new NORGate();
        private ANDGate outputAND = new ANDGate();
        private ANDGate loopAND = new ANDGate();
        private ORGate registerWriteOR0 = new ORGate();
        private ORGate registerWriteOR1 = new ORGate();
        private ORGate registerWriteOR2 = new ORGate();
        private ANDGate registerWriteAND = new ANDGate();
        private Connector[] operation = new Connector[4];
```

```
private Connector clock = new Connector();

public Controller()
{
    for (int index = 0; index < readValueORs.Length; index++)
    {
        readValueORs[index] = new ORGate();
        source.ConnectTo(readValueORs[index].Source);
    }

    // Leftmost column of OR gates
    readValueORs[0].Output.ConnectTo(readValueORs[8].Input0);
    readValueORs[1].Output.ConnectTo(readValueORs[8].Input1);
    readValueORs[2].Output.ConnectTo(readValueORs[9].Input0);
    readValueORs[3].Output.ConnectTo(readValueORs[9].Input1);
    readValueORs[4].Output.ConnectTo(readValueORs[10].Input0);
    readValueORs[5].Output.ConnectTo(readValueORs[10].Input1);
    readValueORs[6].Output.ConnectTo(readValueORs[11].Input0);
    readValueORs[7].Output.ConnectTo(readValueORs[11].Input1);

    // Second leftmost column of OR gates
    readValueORs[8].Output.ConnectTo(readValueORs[12].Input0);
    readValueORs[9].Output.ConnectTo(readValueORs[12].Input1);
    readValueORs[10].Output.ConnectTo(readValueORs[13].Input0);
    readValueORs[11].Output.ConnectTo(readValueORs[13].Input1);

    // Second rightmost column or OR gates
    readValueORs[12].Output.ConnectTo(readValueORs[14].Input0);
    readValueORs[13].Output.ConnectTo(readValueORs[14].Input1);

    // Rightmost column of OR gates
    readValueORs[14].Output.ConnectTo(loopAND.Input1);

    for (int index = 0; index < operation.Length; index++)
    {
        operation[index] = new Connector();
    }

    operation[0].ConnectTo(operationOR0.Input0);
    operation[1].ConnectTo(operationOR0.Input1);
    operation[2].ConnectTo(operationOR1.Input0);
    operation[3].ConnectTo(operationOR1.Input1);

    operation[0].ConnectTo(registerWriteOR0.Input0);
    operation[1].ConnectTo(registerWriteOR0.Input1);
    registerWriteOR0.Output.ConnectTo(registerWriteOR1.Input0);
    operation[2].ConnectTo(registerWriteOR1.Input1);
    registerWriteOR1.Output.ConnectTo(registerWriteAND.Input0);
    clock.ConnectTo(registerWriteAND.Input1);

    operation[3].ConnectTo(loopAND.Input0);

    operationOR0.Output.ConnectTo(operationNOR.Input0);
    operationOR1.Output.ConnectTo(operationNOR.Input1);

    operationNOR.Output.ConnectTo(outputAND.Input0);
    clock.ConnectTo(outputAND.Input1);

    source.ConnectTo(operationOR0.Source);
    source.ConnectTo(operationOR1.Source);
    source.ConnectTo(operationNOR.Source);
    source.ConnectTo(registerWriteOR0.Source);
    source.ConnectTo(registerWriteOR1.Source);
```

```
            source.ConnectTo(registerWriteOR2.Source);
            source.ConnectTo(outputAND.Source);
            source.ConnectTo(loopAND.Source);
            source.ConnectTo(registerWriteAND.Source);
        }

        public Connector Source
        {
            get { return source; }
        }

        public Connector[] Operation
        {
            get { return operation; }
        }

        public Connector Clock
        {
            get { return clock; }
        }

        public Connector[] ReadValue0
        {
            get
            {
                // Each of the leftmost OR gates in the sixteen-input
                // stack of OR gates accepts two of the inputs.
                Connector[] connectors = new Connector[16];
                for (int index = 0; index < connectors.Length / 2; index++)
                {
                    connectors[index * 2] = readValueORs[index].Input0;
                    connectors[index * 2 + 1] = readValueORs[index].Input1;
                }
                return connectors;
            }
        }

        public Connector RegisterWrite
        {
            get { return registerWriteAND.Output; }
        }

        public Connector[] RegisterWriteSource
        {
            get { return new Connector[] { operation[2], operation[1] }; }
        }

        public Connector NextInstructionSource
        {
            get { return loopAND.Output; }
        }

        public Connector Output
        {
            get { return outputAND.Output; }
        }
    }
}
```

17.2 CPU

The main part of the CPU incorporates all of the devices from the previous chapter.

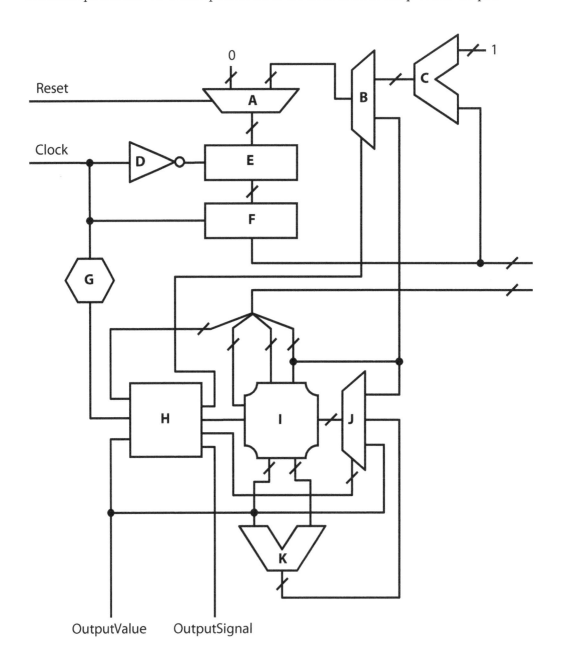

For easy reference, the following table maps the parts in the schematic to the variables that represent them in the implementation.

Device	Variable
A	nextProgramCounter
B	nextInstructionAddress
C	nextInstructionAdder
D	invertedClock
E	programCounter
F	programCounterBuffer
G	delay
H	controller
I	registers
J	registerWriteValue
K	adder

```
using System;
namespace Hardware
{
    class CPU
    {
        private Register programCounter = new Register(4);
        private Register programCounterBuffer = new Register(4);
        private Multiplexer nextProgramCounter = new Multiplexer(4, 1);
        private Multiplexer nextInstructionAddress =
                                        new Multiplexer(4, 1);
        private Adder nextInstructionAdder = new Adder(4);
        private NOTGate invertedClock = new NOTGate();
        private NOTGate zero = new NOTGate();
        private Delay delay = new Delay(1000);
        private Controller controller = new Controller();
        private MemoryBank registers = new MemoryBank(16, 4);
        private Multiplexer registerWriteValue = new Multiplexer(16, 2);
        private Adder adder = new Adder(16);

        private Connector source = new Connector();
```

```
private Connector clock = new Connector();
private Connector reset = new Connector();

public CPU()
{
    // Connect everything that uses sixteen connectors
    // (data carriers)
    for (int index = 0; index < 16; index++)
    {
        registerWriteValue.Output[index].ConnectTo(
            registers.WriteValue[index]);

        registers.ReadAddress1[Math.Min(index, 3)].ConnectTo(
            registerWriteValue.Inputs[0][index]);

        registers.ReadValue0[index].ConnectTo(adder.Input0[index]);
        registers.ReadValue0[index].ConnectTo(
            controller.ReadValue0[index]);
        registers.ReadValue0[index].ConnectTo(
            registerWriteValue.Inputs[2][index]);
        registers.ReadValue1[index].ConnectTo(adder.Input1[index]);

        adder.Output[index].ConnectTo(
            registerWriteValue.Inputs[1][index]);
    }

    // Connect everything that uses four connectors
    // (address carriers)
    for (int index = 0; index < 4; index++)
    {
        zero.Output.ConnectTo(nextProgramCounter.Inputs[1][index]);

        registers.ReadAddress1[index].ConnectTo(
            nextInstructionAddress.Inputs[1][index]);

        programCounter.Output[index].ConnectTo(
            programCounterBuffer.Input[index]);
        programCounterBuffer.Output[index].ConnectTo(
            nextInstructionAdder.Input1[index]);

        nextInstructionAdder.Output[index].ConnectTo(
            nextInstructionAddress.Inputs[0][index]);
        nextInstructionAddress.Output[index].ConnectTo(
            nextProgramCounter.Inputs[0][index]);
        nextProgramCounter.Output[index].ConnectTo(
            programCounter.Input[index]);
    }

    // Create a four-bit binary "one"
    source.ConnectTo(nextInstructionAdder.Input0[0]);
    zero.Output.ConnectTo(nextInstructionAdder.Input0[1]);
    zero.Output.ConnectTo(nextInstructionAdder.Input0[2]);
    zero.Output.ConnectTo(nextInstructionAdder.Input0[3]);

    // Connect all of the clock signals
    invertedClock.Output.ConnectTo(programCounter.Write);
    clock.ConnectTo(invertedClock.Input);
    clock.ConnectTo(programCounterBuffer.Write);
    clock.ConnectTo(delay.Input);
    delay.Output.ConnectTo(controller.Clock);

    // Connect the reset line
    reset.ConnectTo(nextProgramCounter.Selector[0]);
```

```
        // Connect the controller
        controller.NextInstructionSource.ConnectTo(
            nextInstructionAddress.Selector[0]);
        controller.RegisterWrite.ConnectTo(registers.Write);
        controller.RegisterWriteSource[0].ConnectTo(
            registerWriteValue.Selector[0]);
        controller.RegisterWriteSource[1].ConnectTo(
            registerWriteValue.Selector[1]);

        source.ConnectTo(zero.Input);

        // Connect all of the power
        source.ConnectTo(programCounterBuffer.Source);
        source.ConnectTo(zero.Source);
        source.ConnectTo(programCounter.Source);
        source.ConnectTo(nextProgramCounter.Source);
        source.ConnectTo(nextInstructionAddress.Source);
        source.ConnectTo(nextInstructionAdder.Source);
        source.ConnectTo(invertedClock.Source);
        source.ConnectTo(controller.Source);
        source.ConnectTo(registers.Source);
        source.ConnectTo(registerWriteValue.Source);
        source.ConnectTo(adder.Source);
        source.ConnectTo(delay.Source);
    }

    public Connector Source
    {
        get { return source; }
    }

    public Connector Clock
    {
        get { return clock; }
    }

    public Connector Reset
    {
        get { return reset; }
    }

    public Connector[] MemoryReadAddress
    {
        get { return programCounterBuffer.Output; }
    }

    public Connector[] MemoryValueRead
    {
        get
        {
            Connector[] connectors = new Connector[16];
            for(int index = 0; index < 4; index++)
            {
                // Bits 0-4 are read address 1
                connectors[index] = registers.ReadAddress1[index];

                // Bits 4-7 are read address 0
                connectors[index + 4] = registers.ReadAddress0[index];

                // Bits 8-11 are the write address
                connectors[index + 8] = registers.WriteAddress[index];
```

```
                // Bits 12-15 are the operation
                connectors[index + 12] = controller.Operation[index];
            }
            return connectors;
        }
    }

    public Connector[] OutputValue
    {
        get { return registers.ReadValue0; }
    }

    public Connector Output
    {
        get { return controller.Output; }
    }
  }
}
```

The CPU is now fully assembled. This concludes the transistor-based part of the implementation. All of the remaining functionality will be implemented directly, as it is only present to support the CPU that we are simulating.

17.3 BIOS

The BIOS in our system is slightly different than a typical BIOS, as it performs a simpler task. This BIOS is only responsible for obtaining the initial program from storage (i.e. a file) and writing it into memory. The file can be read directly, using built-in functionality from the C# runtime. As main memory is implemented using transistors, we will communicate with it via its connectors. The BIOS will need to be connected to the write-address, write-value, and write-enable connectors of the main memory module. The BIOS will later reset the CPU, causing it to start executing the newly loaded instructions. This will require connections to the reset and clock connectors on the CPU.

```
using System;
using System.IO;
using System.Globalization;
using System.Collections.Generic;

namespace Hardware
{
    class BIOS
    {
        private Connector memoryWrite = new Connector();
        private Connector[] memoryAddress = new Connector[4];
        private Connector[] memoryValue = new Connector[16];
        private Connector clock = new Connector();
        private Connector cpuReset = new Connector();

        public BIOS()
        {
```

```
        // Crete all of the connectors on the BIOS
        for (int index = 0; index < memoryAddress.Length; index++)
        {
            memoryAddress[index] = new Connector();
        }

        for (int index = 0; index < memoryValue.Length; index++)
        {
            memoryValue[index] = new Connector();
        }
    }

    public void Reset()
    {
        // Get the important connectors to a known state
        clock.IsHighVoltage = false;
        memoryWrite.IsHighVoltage = false;
        cpuReset.IsHighVoltage = false;

        // Read all of the lines in the instruction file
        string[] fileLines;
        try
        {
            fileLines = File.ReadAllLines("instructions.txt");
        }
        catch (Exception exception)
        {
            Console.Write("Error opening instructions.txt: "
            Console.WriteLine(exception.Message);
            return;
        }

        // Store the valid part of each line in a list
        List<string> instructionStrings = new List<string>();
        for (int index = 0; index < fileLines.Length; index++)
        {
            string line = fileLines[index];

            // See if the line contains a comment
            if (line.Contains(";"))
            {
                // Get the section of the line from position zero
                // to the position where the comment starts
                // thus discarding the comment.
                line = line.Substring(0, line.IndexOf(";"));
            }

            // Remove any spaces/tabs/etc. at the
            // beginning and end of the line
            line = line.Trim();

            // If there is anything left after all of
            // the tabs and spaces are gone, add it to
            // the list of valid strings
            if (line.Length > 0)
            {
                instructionStrings.Add(line);
            }
        }

        // Ensure we don't have more instructions
        // than main memory can store
        if (instructionStrings.Count > 16)
```

```
        {
            Console.WriteLine("Too many instructions!");
            return;
        }

        // Send the instructions to memory
        for (ushort index = 0; index < instructionStrings.Count;
            index++)
        {
            ushort instruction = 0;
            try
            {
                // Convert the sequence of characters in the
                // string into the number they represent
                instruction = ushort.Parse(instructionStrings[index],
                    NumberStyles.HexNumber);
            }
            catch (Exception)
            {
                Console.Write("Couldn't understand instruction ");
                Console.WriteLine("number " + index);
                return;
            }

            // Set the write address on the write address
            // lines of main memory
            SetValue(memoryAddress, index);

            // Set the write value on the write value
            // lines of main memory
            SetValue(memoryValue, instruction);

            // Enable writing so the value is stored
            memoryWrite.IsHighVoltage = true;

            // Disable writing
            memoryWrite.IsHighVoltage = false;
        }

        // Reset the CPU
        cpuReset.IsHighVoltage = true;
        clock.IsHighVoltage = true;
        clock.IsHighVoltage = false;
        cpuReset.IsHighVoltage = false;
    }

    public Connector MemoryWrite
    {
        get { return memoryWrite; }
    }

    public Connector[] MemoryAddress
    {
        get { return memoryAddress; }
    }

    public Connector[] MemoryValue
    {
        get { return memoryValue; }
    }

    public Connector CPUReset
    {
```

```
            get { return cpuReset; }
        }

        public Connector Clock
        {
            get { return clock; }
        }

        private static void SetValue(Connector[] connectors, uint value)
        {
            for (int index = 0; index < connectors.Length; index++)
            {
                connectors[index].IsHighVoltage = ((value & 1) == 1);
                value = value >> 1;
            }
        }
    }
}
```

Each value to load is read from the file named "instructions.txt". Once read, each string of characters is converted to the number that it represents, using the "Parse" method of the "ushort" type.

Setting multi-bit values is made a bit easier with a small helper method. The SetValue method simply sets each connector to high or low voltage based on each bit of a number being a one or a zero. The AND operation isolates the rightmost bit, which is compared to one to see if the voltage should be high voltage or not. The subsequent shift operation moves the next bit into the rightmost position.

The BIOS is now completed. If connected to main memory and the CPU, the BIOS will load instructions from a file into the main memory unit, and then reset the CPU.

17.4 Display

The display is a simple class that accepts the information from the OutputValue and OutputSignal connectors on the CPU. Whenever the output signal switches from low voltage to high voltage, the number on the value connectors should be displayed. The following class accomplishes this task by making use of the existing console window, via the Console.WriteLine() method.

```
using System;

namespace Hardware
{
    class Display
    {
        private Connector printSignal = new Connector();
        private Connector[] printValue = new Connector[16];
```

```
public Display()
{
    for (int index = 0; index < printValue.Length; index++)
    {
        printValue[index] = new Connector();
    }

    printSignal.VoltageChanged += OnPrintSignalChange;
}

public Connector PrintSignal
{
    get { return printSignal; }
}

public Connector[] PrintValue
{
    get { return printValue; }
}

private void OnPrintSignalChange()
{
    if (printSignal.IsHighVoltage)
    {
        Console.WriteLine(GetValue(printValue));
    }
}

private static uint GetValue(Connector[] connectors)
{
    uint value = 0;
    for (int index = connectors.Length - 1; index >= 0; index--)
    {
        value = value << 1;
        if (connectors[index].IsHighVoltage)
        {
            value = value | 1;
        }
    }
    return value;
}
    }
}
```

Aside from printing the current output value, this class contains a helper method to create an integer from the voltages on an array of connectors. The value on each connector is merged into a single integer. The process starts with the last connector, which carries the leftmost bit of the number. As each connector is processed, the integer is shifted to the left one bit, and the next bit is set. The integer contains one bit from each connector when the method is complete.

17.5 Main Program

All of the components have now been created. The only task remaining is to interconnect connect each of the major components: the CPU, main memory, BIOS, display, clock, and power source. As we have already implemented each of these components, we can simply instantiate each one.

```
using System;

namespace Hardware
{
    class Program
    {
        static CPU cpu = new CPU();
        static BIOS bios = new BIOS();
        static Display display = new Display();
        static MemoryBank memory = new MemoryBank(16, 4);

        static Connector highVoltageSource = new Connector();
        static NormalTransistor clock = new NormalTransistor();

        static void Main(string[] args)
        {
            // Connect the CPU for reading from main memory
            for (int index = 0; index < memory.ReadAddress0.Length;
                index++)
            {
                cpu.MemoryReadAddress[index].ConnectTo(
                    memory.ReadAddress0[index]);
            }

            for (int index = 0; index < memory.ReadValue0.Length; index++)
            {
                cpu.MemoryValueRead[index].ConnectTo(
                    memory.ReadValue0[index]);
            }

            // Connect the BIOS for writing to main memory
            for (int index = 0; index < memory.WriteAddress.Length;
                index++)
            {
                bios.MemoryAddress[index].ConnectTo(
                    memory.WriteAddress[index]);
            }

            for (int index = 0; index < memory.WriteValue.Length; index++)
            {
                bios.MemoryValue[index].ConnectTo(
                    memory.WriteValue[index]);
            }

            bios.MemoryWrite.ConnectTo(memory.Write);

            // Connect the CPU reset line
            cpu.Reset.ConnectTo(bios.CPUReset);

            // Connect the CPU to the display unit
            cpu.Output.ConnectTo(display.PrintSignal);
```

```
        for (int index = 0; index < cpu.OutputValue.Length; index++)
        {
            cpu.OutputValue[index].ConnectTo(
                display.PrintValue[index]);
        }

        // Connect the clock
        clock.Source.ConnectTo(highVoltageSource);
        clock.Output.ConnectTo(bios.Clock);
        clock.Output.ConnectTo(cpu.Clock);

        // Connect the power
        cpu.Source.ConnectTo(highVoltageSource);
        memory.Source.ConnectTo(highVoltageSource);

        // Turn on the power
        highVoltageSource.IsHighVoltage = true;

        // Reset the system
        bios.Reset();

        // Run the clock until a key is pressed
        while (!Console.KeyAvailable)
        {
            clock.Input.IsHighVoltage = true;
            clock.Input.IsHighVoltage = false;
        }
    }
  }
}
```

17.6 Exercises

1. The majority of the system is made of individual transistors. Modify the code to count the number of transistors used in then system.

2. A large number of transistors are used to create the main memory that simply stores instructions. Update the transistor-counting code to count all transistors except those used in the main memory.

Moving Forward

Many topics were briefly introduced throughout this book. Computer-related development contains a great number of disciplines that you may wish to investigate further.

Physical hardware design is typically the most expensive discipline, but several devices have been created to reduce the cost and complexity. Hardware devices known as CPLDs and FPGAs are available from a variety of companies, and allow for quick, cost-effective design. After designing the desired schematics on a computer, these devices can be used to synthesize a physical implementation of the schematic. This device may then be connected to other electronic peripherals with which it can interact. FPGAs and CPLDs are often sold on a development board, with many useful peripherals already attached.

If the direct hardware programming was of interest, you can easily extend the sample operating system, or create one of your own. With a few simple additions, such as keyboard input and disk access, the simple operating system can perform many useful tasks. The sample code contained in the resources bundle includes a 64-bit version of the system that may be a more appropriate starting point for a new operating system. Creating a simple text editor and assembler will allow you to program your operating system from within itself.

Finally, you may find application programming to be the most appealing. This is typically the most popular choice, as applications can be created with more powerful tools, and can be distributed to other users with ease. Free development tools are available for C# and other programming languages. There are many more features of the C# language that you may wish to discover, as well as a plethora of classes available in the runtime library.

Of course, all three areas may be combined into a single system. If you are feeling ambitions, you may wish to try developing a simple microprocessor, along with a small operating system and application suite to run upon it. With a bit of creativity, programmable computer development can lead to an endless supply of challenging and rewarding projects.

Answers

These following sections contain answers for the exercises at the end of each chapter. Note that there are other valid answers for the majority of questions.

Chapter 2

1. $(15)_{10} = (1111)_2$
   $(7)_{10} = (111)_2$
   $(127)_{10} = (1111111)_2$
   $(63)_{10} = (111111)_2$

   They are all strings of ones because they are one less than a power of two.

2. $(1101 + 0011) = 0000$
   $(1010 + 0101) = 1111$
   $(1111 + 0001) = 0000$

3. There is no negative zero in two's complement. Positive zero is $(0000)_2$. Inverting gives $(1111)_2$. Adding one gives $(0000)_2$, which is the same as positive zero. There is only one zero.

4. A two's complement number is negative if the leftmost bit is a one. All hexadecimal digits higher than seven have a leftmost bit of one. If a hexadecimal number's leftmost digit is 8, 9, A, B, C, D, E or F, the number is negative.

Chapter 3

1.

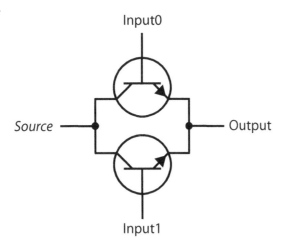

Input0

Source ———•   •——— Output

Input1

2. The output is high voltage when either of the inputs are high voltage.

3.

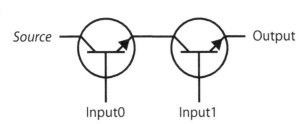

Source ———   ——— Output

Input0   Input1

4. The output is high voltage when both inputs are high voltage.

Chapter 4

1.

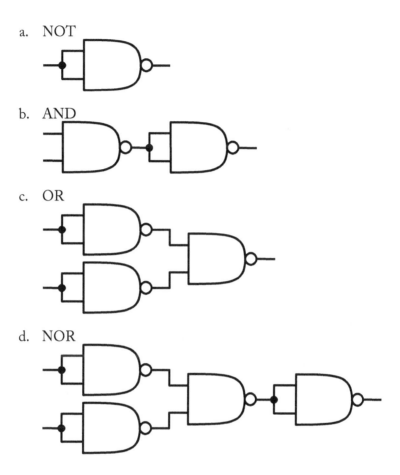

a. NOT

b. AND

c. OR

d. NOR

Using only NAND gates, we can create NOT, AND, NAND, OR, and NOR gates. We created XOR gates from NAND, OR, and AND gates, and we can now create those from NAND gates. This means that with only NAND gates, all other gates can be created. This is also true of NOR gates.

2.

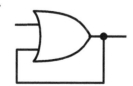

3.

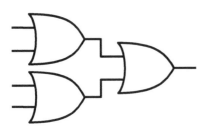

4.

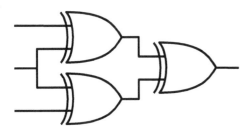

Chapter 5

1. The register stores 5 at the end of each sequence. Registers always output the value they are storing, so they also both output 5.

2. Thirty-two two-input multiplexers can accept sixty-four inputs and reduce them to thirty two outputs. Sixteen more multiplexers will accept those thirty-two as inputs and reduce them to sixteen outputs, and so on. In total, 63 are needed (32 + 16 + 8 + 4 + 2 + 1 = 63).

Chapter 6

1.

   a. These instructions simply loop forever. The first stores a non-zero value in register zero, and the second jumps to the first if there is a non-zero vale in register zero.

   b. There is no "end of program" indicator in our assembly language. Without this infinite loop at the end of the program, the CPU will simply continue to read the next value in memory, whatever it may be, and interpret it as an instruction.

   Each memory location always contains a value, even if one hasn't been manually set (each wire will have either high voltage or low voltage). When a device is powered on, registers may initially contain all zeroes, all ones, or a seemingly random mix of values. You can read from registers (or memory) before ever writing a value there, you just have no guarantee of what value you will receive. Feel free to try writing a program that does this. Simply use the "print" operation to print the contents of a register you have never written to.

2. Multiplication can be achieved by starting with zero, then repeatedly adding one of the two numbers to be multiplied. The other number specifies how many times it should be added. For example, five multiplied by six can be calculated by starting with zero, then adding five repeatedly (six times). Alternately, you could start with zero and add six, five times.

   The following code performs multiplication in this manner. The result (register two) is initialized to zero, then the second number (register one) is added to the result repeatedly. The first number (register zero) is decremented each time, acting as a counter.

```
set #0, 0x5      ; store the first number to be multiplied
set #1, 0x6      ; store the second number to be multiplied
set #2, 0x0      ; set the result to 0
set #3, 0xF      ; set register3 to -1
add #4, #1, #2   ; add the second number to the result
copy #2, #4      ; copy the sum back into the result register
add #4, #0, #4   ; decrease the counter
copy #0, #4      ; copy the sum back into the counter register
loop, #0, 0x4    ; if the counter isn't 0 yet, keep going
print #2         ; print the result
set #0, 0x1      ; make sure the loop is taken
loop #0, 0x0     ; do it all again
```

3.

```
1005    ; store the first number to be multiplied
1106    ; store the second number to be multiplied
1200    ; set the result to 0
130F    ; set register3 to -1
4412    ; add the second number to the result
2240    ; copy the sum back into the result register
4403    ; decrease the counter
2040    ; copy the sum back into the counter register
8004    ; if the counter isn't 0 yet, keep going
0020    ; print the result
1001    ; make sure the loop is taken
8000    ; do it all again
```

Chapter 7

1. This implementation of the CPU can accept copy instructions with the same register used for both reading and writing. This instruction will connect the specified register's output to its input, which stays in a steady state. This has the same effect as "copying" the value from a register back into itself.

2. This CPU cannot support adding into one of the registers being read from. This situation would connect the output of a register to the adder, and the output of the adder to the register's input. This arrangement will never reach a steady state.

    Whenever a value is written to the register, it will immediately appear on the output of the register, causing it to be added again, causing a new value to be written, which will be added again, and so on, until the write signal is disabled to break the loop. An unexpected value would be in the register when the write signal was disabled, not the expected sum.

Chapter 10

1.

```
; Store current position
mov dword [0x100000], 0xB8000

push 1
call ShowNumber
push 2
call ShowNumber
push 3
call ShowNumber
push 4
call ShowNumber
push 5
call ShowNumber

; wait forever
jmp $


ShowNumber:
push ecx              ; Save ECX, as we overwrite it as well

push ebx              ; The stack now contains the value that was in EBX,
                      ; followed by the value that was in ECX,
                      ; followed by the return address pushed by "call".

mov bl, [esp + 12]    ; Copy the number to BL. The number is 12 bytes down
                      ; the stack, after the 4-byte EBX, 4-byte ECX, and
                      ; the 4-byte return address.

add bl, '0'           ; Adding the ASCII character zero will convert the
                      ; number to a corresponding ASCII character.
                      ; e.g. the number 5
                      ; '0' = 0x30. 0x30 + 5 = 0x35 0x35 = '5'
                      ; See the ASCII table for more details.

mov ecx, [0x100000]   ; Load the current position from memory

mov [ecx], bl         ; Display the character at the current position

add dword [0x100000], 2   ; Move the current position forward
                          ; 1 character (which is two bytes, one for the
                          ; symbol, one for the colors)

pop ebx
pop ecx               ; Restore the value of EBX and ECX to whatever
                      ; they were before this function was called. The
                      ; return address is now the top item on the stack

ret                   ; Return to the address on the top of the stack
```

2.

```
push 0xC
call ShowNumber

; wait forever
jmp $


ShowNumber:
push ebx                    ; The stack now contains the value that was in EBX,
                            ; followed by the return address pushed by "call".

mov bl, [esp + 8]           ; Copy the number to BL. The number is 8 bytes down
                            ; the stack, after the 4-byte EBX, and the 4-byte
                            ; return address.

cmp bl, 9                   ; See if the number is bigger than 9
jg letter                   ; If greater than nine, we need to print a letter

add bl, '0'                 ; Adding the ASCII character zero will convert the
                            ; number to a corresponding ASCII character.
                            ; e.g. the number 5
                            ; '0' = 0x30. 0x30 + 5 = 0x35 0x35 = '5'
                            ; See the ASCII table for more details.
jmp print

letter:
sub bl, 10                  ; Subtract 10, so 10 goes to 0, 11 goes to 1, etc.
add bl, 'A'                 ; Add to 'A', 'A' + 0 = 'A', 'A' + 1 = 'B'.
                            ; See the ASCII table for more details.


print:
mov [0xB8000], bl           ; Display the character

pop ebx                     ; Restore the value of EBX to whatever it was
                            ; before this function was called. The return
                            ; address is now the top item on the stack

ret                         ; Return to the address on the top of the stack
```

Chapter 11

1. Each column of a binary number represents a value twice as large as the column to its immediate right. Hence, shifting all of the bits to the right one space will make each one represent a number half as large. If each digit represents a value that is half as large as it used to be, the entire number will be half as large. In short, shifting the binary number to the right one digit will have the same effect as dividing by two. In x86 assembly, the "shr" instruction will shift a number to the right a specified number of digits.

2. Consider the decimal number 42,123. When dividing by one thousand, the remainder is simply all digits to the right of the thousands digit; the "42 thousand" part of the number is evenly divided by one thousand, while the digits to the right of the thousands digit – "123" – are smaller than one thousand and are left as a remainder.

   The same pattern applies to binary digits. The number eight is simply bit number three set to a one: $(00001000)_2$. When dividing by eight, the eights column and all columns to the left can be evenly divided by eight, and all digits to the right of the eights column are the remainder. E.g. $(11111101)_2$ divided by $(00001000)_2$ has a remainder of $(101)_2$. In decimal, that is $253 \div 8$ has a remainder of 5. Therefore, the remainder of a division by eight can be found by isolating the rightmost three bits. This can be accomplished with an AND operation using an argument of $(00000111)_2$. The leftmost five bits will be removed (anything AND zero is zero), and the rightmost three bits will be preserved (zero AND one is zero, one AND one is one).

Chapter 12

1. One program can be given more or less time to run by setting the countdown (in the timer handler) to a different value based on which program is being run.

2. The task switching in the sample operating system stops one program from running and allows the other program to run. This is not the same as switching which window is in focus in a graphical operating system. Even when a window is hidden, the operating system continues to give it a chance to run. You can deduce this from the fact that even hidden programs can carry out operations (performing a large copy operation, receiving messages from the internet, etc.).

3. The operating system could run multiple instance of the same application. Each application has its own instruction pointer value, so the two instances could run independently (each at a different part of the program at any given time). Running two of the sample application, e.g. two of Program A, would allows one instance to overwrite the display output of the other, as they both write to the same area of the screen.

Chapter 13

1. An array of fifty-seven items contains items numbered zero through fifty-six. The largest valid index is fifty-six.

2. The value within the parenthesis must be a Boolean value (true or false). Other types will generate an error, as they do not make sense.

3. The first step is to isolate bit number five. The number 32 has a binary representation of all zeroes, except for bit number five. Performing an AND operation with the value 32 will isolate bit five. With bit five as the only bit that could be set, determining if it is set can be accomplished by comparing the result to zero. If a number AND 32 is zero, bit number five was not set to one. If a number AND 32 is not zero, bit number five was set to one.

Chapter 14

1.

```
void Print(int counter)
{
    if (counter == 0)
    {
        // Printed 100 times, stop
        return;
    }
    else if (counter == -1)
    {
        // First call, start printing
        Print(100);
    }
    else
    {
        // counter is somewhere from
        // 100 to 1, print hello.
        System.Console.WriteLine("Hello");
    }
}
```

2. Exceptions are only caught by travelling back through the stack of calls until a "try" block is found. Once an exception is caught, the program exits the "try" block and begins executing the "catch" block to handle the caught exception. Because the point of execution is no longer within that "try" block, simply throwing the exception that was just caught will cause it to continue to the next "try" block it finds.

3. The "using" statement indicates which namespaces should be checked for classes that are used (to save you the trouble of typing the namespace every time). Console is a class within the System namespace; it is not a namespace, so it cannot be included in a "using" statement.

Chapter 15

1. The queue class adds new items to the end of the group and removes items from the start of the group. The list class adds items to the end of the group, so the only thing that needs to be done is to always remove items from the start of the group. The start item can be accessed at index zero:

```
firstItem = myList[0];
```

And the item can be removed using the RemoveAt function:

```
myList.RemoveAt(0);
```

2. The infinite loop is a valid concern. To prevent a never-ending loop, the "set" accessor checks to see if the value has actually changed before doing anything further. If the value is the same, the accessor doesn't continue propagating the voltage to other connectors. This ensures that whenever the voltage does change, each connector is only updated once, rather than repeatedly sending and receiving the same value.

   If connector1 changes to high voltage, it will notify each connected connector (e.g. connector2). Connector2 would then notify each connector that it is connected to, which would include connector1. Because connector1 is already at the new voltage, the "set" accessor will exit immediately without creating a redundant notification.

3. The presented design is not the fastest way to simulate the instructions. Rather than keeping track of the thousands of connectors and transistors, the program could simply interpret the instructions directly.

   For example, if the current instruction was an "add" operation, the C# program could just add the two values using the "+" operator, rather than simulating an adder's hardware. However, simulating each transistor is useful for validating and debugging the hardware design.

Chapter 17

1. To count each transistor we need one counter variable for all of the transistors. Whenever a transistor is created, the counter should be incremented. The constructor is run whenever a class is created, so the transistor constructor is an ideal place to increment the counter.

    In the "Program" class we can create the one counter for all transistors, as well as print its value. When the Program class is instantiated, it will have space reserved for each member variable. As each transistor is created, the value in the space for the "transistorCount" variable is incremented. When the program class has been fully instantiated, the Main method runs, printing the total transistor count.

```
class Program
{
    public static int transistorCount;

    static CPU cpu = new CPU();
    static BIOS bios = new BIOS();
    static Display display = new Display();
    static MemoryBank memory = new MemoryBank(16, 4);

    private static Connector highVoltageSource = new Connector();
    private static NormalTransistor clock = new NormalTransistor();

    static void Main(string[] args)
    {
        Console.WriteLine("Transistor count: " + transistorCount);

... (rest of the Program class omitted)
```

Now we just need to make each transistor increment the counter.

```
class NormalTransistor
{
    private Connector source = new Connector();
    private Connector input = new Connector();
    private Connector output = new Connector();

    public NormalTransistor()
    {
        source.VoltageChanged += UpdateOutput;
        input.VoltageChanged += UpdateOutput;
        UpdateOutput();
        Program.transistorCount++;
    }

... (rest of the NormalTransistor class omitted)
```

```
class InvertingTransistor
{
    private Connector source = new Connector();
    private Connector input = new Connector();
    private Connector output = new Connector();

    public InvertingTransistor()
    {
        source.VoltageChanged += UpdateOutput;
        input.VoltageChanged += UpdateOutput;
        UpdateOutput();
        Program.transistorCount++;
    }
}
... (rest of the InvertingTransistor class omitted)
```

Building and running the CPU emulator produces the result:

"Transistor count: 12833".

2. Obtaining the transistor count without the main memory included can be accomplished by moving the creation of main memory to a location after the transistor counter is displayed.

```
class Program
{
    public static int transistorCount;

    static CPU cpu = new CPU();
    static BIOS bios = new BIOS();
    static Display display = new Display();
    static MemoryBank memory;

    private static Connector highVoltageSource = new Connector();
    private static NormalTransistor clock = new NormalTransistor();

    static void Main(string[] args)
    {
        Console.WriteLine("Transistor count: " + transistorCount);

        memory = new MemoryBank(16, 4);
```
... (rest of the InvertingTransistor class omitted)

This yields:

"Transistor count: 7350"

Index

www.ingramcontent.com/pod-product-compliance
Lightning Source LLC
Chambersburg PA
CBHW080352060326
40689CB00019B/3986